CONVOLUTION AND FOURIER TRANSFORMS *for* COMMUNICATIONS ENGINEERS

CONVOLUTION AND FOURIER TRANSFORMS

for

COMMUNICATIONS ENGINEERS

R. D. A. Maurice, *OBE, Dr.-Ing, Ing ESE, C Eng, FIEE, FRTS*

BBC Engineering Division

PENTECH PRESS
London

First Published, 1976
by Pentech Press Limited
London: 8 John Street, WC1N 2HY

ISBN 0 7273 0301 5

PRINTED BY Unwin Brothers Limited
THE GRESHAM PRESS OLD WOKING SURREY ENGLAND

Produced by 'Uneoprint'

A member of the Staples Printing Group

Contents

Preface

My experience of learning mathematics at schools and colleges in England, California and France was that, whether my teachers were lovers of mathematics or not, at least they all seemed to be very good at the subject. I can remember staying after school in the classroom of an American high school (a comprehensive school of 2200 pupils) one day, because I refused to admit to understanding that the number of spaces between n telegraph poles was n-1, and I hadn't the sense, at the time, to ask whether the line of poles had two ends to it or whether it formed a closed loop! In any case, I should have been told that (in the 1920s) ring-main communication circuits were highly unusual. I hope the reader will forgive me, a non-mathematician, for having the temerity to write a book on one or two aspects of mathematics; in particular, some of those that have a fairly close bearing on some aspects of electronic engineering; in particular, to broadcasting engineering. Anyway, whether I'm forgiven or not, no one can take away the very great pleasure I have had in clarifying my thoughts and writing them down. I have taken pains to ensure that the steps that lead along the mathematical paths followed in this book are properly explained and I have avoided like the plague the temptation to look clever by omitting essential reasoning.

There are enough elementary statistics in the early chapters to warn the reader that statistics tell one nothing that one didn't know before, but that they are a great aid to the orderly presentation of facts; furthermore, the introduction of the philosophy of convolution seems to me to be easier for the student to grasp if a statistical approach is adopted, notwithstanding the fact that convolution is becoming more and more widely applied to electric circuit problems that are not, macroscopically at any rate, statistical.

The treatment of the behaviour of cascaded circuits by convolution is merely an alternative to the method that uses the multiplication of Fourier transforms or Laplace transforms, and so this book contains several applications or practical examples in which both methods are demonstrated.

I have made no attempt to flood the book with examples, but where they are given, they are made to conform closely to the real practical situation and they are therefore not necessarily simple. The study of the spectrum of a television signal is, perhaps, an example that illustrates this point.

The delta function or unit impulse is given pride of place as a test signal as well as playing the role of a time shifter. It is, in my view, a very powerful tool whose properties, although relatively easy to understand, are too little used by engineers.

Convolution or serial division and 'square-rooting' are mentioned, although they are little used, in practice.

A final chapter on correlation is included, merely to show that it is not the same process as convolution, although at first glance it looks like it.

The only mathematical knowledge required for the understanding of this book is that required for the Advanced Level syllabus in pure mathematics or pure and applied mathematics or mathematics, depending upon which examining-board syllabus is assumed.

My grateful thanks are due to Eileen Tasker who, in gaps between her other work, has set down in orderly and neat form my much-corrected manuscript.

R. D. A. Maurice

1. Introduction

Let us start by examining a long multiplication; say 123 × 217:

Table 1.1

Tens of Thousands	Thousands	Hundreds	Tens	Units
		1	2	3
		2	1	7
		8	6	1
	1	2	3	0
2	4	6	0	0
2	6	6	9	1

We have, quite unconsciously, used the decimal system, and 'carrying' from one column to the next on the left presented no problem. Now let's try it in binary notation.

First

$$123 = 100 + 20 + 3$$
$$= (2^6 + 2^5 + 2^2) + (2^4 + 2^2) + (2^1 + 2^0)$$
$$= 2^6 + 2^5 + 2^4 + 2^3 + 2^1 + 2^0 \text{ because } 2 \times 2^2 = 2^3$$

and

$$217 = 200 + 10 + 7$$
$$= (2^7 + 2^6 + 2^3) + (2^3 + 2^1) + (2^2 + 2^1 + 2^0)$$
$$= 2^7 + 2^6 + 2^4 + 2^3 + 2^0$$

because $2^0 + 2 \times 2^1 + 2^2 + 2 \times 2^3 + 2^6 + 2^7$

$$= 2^0 + 2 \times 2^2 + 2 \times 2^3 + 2^6 + 2^7$$
$$= 2^0 + 3 \times 2^3 + 2^6 + 2^7$$
$$= 2^0 + 2^3 + 2^4 + 2^6 + 2^7$$

Table 1.2

	2^{14}	2^{13}	2^{12}	2^{11}	2^{10}	2^{9}	2^{8}	2^{7}	2^{6}	2^{5}	2^{4}	2^{3}	2^{2}	2^{1}	2^{0}
									1	1	1	1	0	1	1
								1	1	0	1	1	0	0	1
									1	1	1	1	0	1	1
								0	0	0	0	0	0	0	0
							0	0	0	0	0	0	0	0	0
						1	1	1	1	0	1	1	0	0	0
					1	1	1	1	0	1	1	0	0	0	0
				0	0	0	0	0	0	0	0	0	0	0	0
			1	1	1	1	0	1	1	0	0	0	0	0	0
		1	1	1	1	0	1	1	0	0	0	0	0	0	0
		1	1+1	1+1	1+1+1	1+1+1	1+1+1	1+1+1+1	1+1+1	1+1	1+1+1	1+1	0	1	1
Carry to the appropriate left-hand column:	0	0	2^{14}	2^{12}	2^{12}	2^{11}	2^{10}	2^{9}	2^{8}	2^{6}	2^{6}	2^{4}	0	0	0
Left in given column after carrying has been done:	1	1	0	1	0	0	0	0	1	0	0	0	0	1	1

Now we indicate whether we have (by a 1) or whether we have not (by a 0) a 2^x in the column of 2^xs, thus we arrive at Table 1.2.

Now, the carrying problem is a tedious one and the penultimate line shows in each column what we have to carry upwards (leftwards). The last line contains the remainders in each column after the carrying has been accomplished. The reader can check that

$$2^{14} + 2^{13} + 2^{11} + 2^6 + 2^1 + 2^0 = 26\,691$$

1.1 ELEMENTARY PROBABILITY

Now suppose we don't wish to subdivide our numbers into columns of units, tens, hundreds or ones, twos, fours, eights, etc., but that we have columns to signify something quite different. For example, suppose we have an urn containing a very large number of red, green and blue balls in which the proportion of red balls is r, green g, and blue b. Since there are only red, green and blue balls, we must have

$$r + g + b = 1$$

Now suppose we shut our eyes and select a ball. The chances that it will be red are r, green g and blue b. Suppose now that we make a further blindfold selection, what combinations are we likely to get? Let's square $r + g + b = 1$

Table 1.3

r	$+g$	$+b$			
r	$+g$	$+b$			
r^2	$+rg$	$+rb$			
	$+rg$		$+g^2$	$+gb$	
		$+rb$		$+gb$	$+b^2$
r^2	$+ 2rg$	$+ 2rb$	$+ g^2$	$+ 2gb$	$+ b^2$

This tells us that the likelihood of getting two red balls is r^2, of a red and a green, $2rg$; of a red and a blue, $2rb$; of two greens, g^2, and so on. Why does it tell this? Well, suppose $r = g = b = 1/3$; then the chances of getting various combinations of two coloured balls are as follows:

Table 1.4

Coloured balls	Probabilities
2 reds	1/9
2 greens	1/9

Table 1.4 continued

2 blues	1/9
red and green	2/9
red and blue	2/9
green and blue	2/9

Obviously there are twice as many chances of getting two different colours as of getting the same colour twice, because the first selection of a pair of different colours can be either one of the two colours, and the probability of picking either green or red is twice the probability of picking, say, only green. This is an application of the 'addition rule' of probability theory, which states that the probability for the occurrence of any one of a number of mutually exclusive events is the sum of the probabilities of occurrence of each event. The 'multiplication rule' states that the probability of occurrence of all of a given number of mutually exclusive events is the product of the probabilities of occurrence of each event. Very simply, consider tossing a statistically balanced coin. The probability for a head is $\frac{1}{2}$; the probability for two heads (either simultaneously by tossing two coins or sequentially by tossing one coin twice) is $\frac{1}{2} \times \frac{1}{2} = \frac{1}{4}$. The probability for *either* a head or a tail is $\frac{1}{2} + \frac{1}{2} = 1$.

Returning now to our combinations of coloured balls, we see that the columns implicit in Table 1.3 are quite different from those in Tables 1.1 and 1.2, yet columns there are and they contain important information, but it isn't concerned with the mechanics (one might say) of multiplication. Table 1.3 assumes that we know how to multiply and it is so constructed as to distinguish the columns according to the colour(s) of balls whose probabilities of being selected are the numbers in the table. Let us set out, *a priori*, the format of a table which would be suitable for recording the information in Table 1.3.

Table 1.5

1 *red*	1 *green*	1 *blue*	2 *reds*	2 *greens*	2 *blues*	*red* + *green*	*red* + *blue*	*green* + *blue*
r	g	b						
r	g	b						
			r^2	g^2	b^2	$2rg$	$2rb$	$2gb$

The product of the two sets of probabilities in the first two lines of the first three columns is given in the third and last lines of Table

1.5. This is a different use of columns from that used in ordinary multiplication. If we had made three selections from our urn of coloured balls we would have cubed our expression $r + g + b$ and the relevant columns would have been headed as follows:

Table 1.6

3R	3G	3B	2R + 1G	2R + 1B	2G + 1R	2G + 1B	2B + 1R	2B + 1G	1R+1G+1B
r^3	g^3	b^3	$3r^2g$	$3r^2b$	$3rg^2$	$3g^2b$	$3rb^2$	$3gb^2$	$6rgb$

where R = red G = green B = blue. To aid thinking, let $r = g = b = 1/3$ again.

1.2 PERMUTATIONS WITH REPETITION

It is not quite obvious that the probability of selecting two balls of one colour and one of a second colour, say $3r^2g$, is three times the probability of selecting three balls all of the same colour.

First we write

$$3r^2g = 3r^3 \text{ because } g = r$$

but, r^3 is the probability of selecting three reds. To make the matter clear we note, first, that r^2 is the probability of selecting two reds, g the probability of selecting one green, so r^2g is the probability of selecting, in sequence, two reds and a green. I say 'in sequence' because it's easier to think of in that way, but the result will apply for a simultaneous selection, in one handful so to speak. It's like tossing either one coin twice or two coins simultaneously. So far, so good, but we might have selected one red, one green and one red, so that's a second way of obtaining two reds and a green and finally we might have selected a green and two reds, so there we have our three ways of obtaining two reds and a green—a combination in which the order of choice doesn't matter. The addition rule of probability comes in once again, because we count as a successful selection of three balls, one in which we have either red, red, green; or red, green red; or green, red, red. This is in fact, the number of permutations of three things of which one is repeated once. The number of permutations of m things, all different, is $m!$ If there are some of the m things that are repeated a times and others b times and yet others c times, and so on, then the number of permutations with repetition is

$m!/a!b!c!$ and so on.

Thus, a combination of two red balls and one green ball is made up of all the permutations of three balls of which two (the red) are alike;

so the probability of getting a handful of three balls consisting of two reds and a green is

$$3!/2! = 3$$

The probability of picking a combination consisting of one red, one green and one blue—the last column of Table 1.6—is

$$3! = 6$$

because we are not allowing for any repetitions in this case. If we made a selection of four balls we should raise $r + g + b$ (=1) to the fourth power and the table, like Table 1.6, would be as follows:

Table 1.7

4R	4G	4B	1R+3G	1R+3B	3R+1G	3R+1B	1G+3B	1B+3G
r^4	g^4	b^4	$4rg^3$	$4rb^3$	$4r^3g$	$4r^3b$	$4gb^3$	$4g^3b$

2R+2G	2R+2B	2G+2B	1R+1B+2G	1R+1G+2B	1G+1B+2R
$6r^2g^2$	$6r^2b^2$	$6g^2b^2$	$12rbg^2$	$12rgb^2$	$12gbr^2$

The coefficient, 4, in front of the (3 + 1) combinations is, of course, the number of permutations of four things with one of them repeated three times, that is

$$4!/3! = 4$$

The coefficient for the (2 + 2) combination is

$$4!/(2! \times 2!) = 6$$

and the coefficient for the (1 + 1 + 2) combinations is

$$4!/2! = 12$$

We can now generalise a little and suppose we have seven different colours of coloured balls: red, green, blue, cyan (C), magenta (M), orange (O), yellow (Y). The probability of selecting the combination of eleven balls as follows

$$3R + 1G + 2B + 1C + 1M + 2O + 1Y$$

will be $[11!/(3!\,2!\,2!)]\ r^3gb^2cmo^2y$

or 1 665 000 $r^3gb^2cmo^2y$

where the small letters represent the proportions of each colour in the infinite urn of coloured balls. Of course,

$$r + g + b + c + m + o + y = 1$$

If the reader now says to himself 'the only difference between Table 1.5 and Table 1.3 is the omission of the + signs from the former that were in the latter', I reply that the + signs in Table 1.3 have no real meaning except to remind us that the sum of $r^2 + 2rg + 2rb + g^2 + 2gb + b^2$ is one; which we knew already, since in squaring $r + g + b$ we were squaring one. True, the + signs allow us to add the various probabilities, but you can't add red to green or green to blue (except when dealing with colorimetry, which is not the subject of the present discussion). Of course, you could add the weights of the balls just as you can add their probabilities, so one could have made a 'weights' table, but this would not have involved multiplication as we had to do in order to find the probabilities of selection of various combinations of coloured balls.

Tables 1.3 to 1.7 are examples of convolution. They are tools for finding all possible combinations that can be formed from two or more groups of things. The 'things' can be colours of coloured balls, statistical errors that are acquired by some force function such as voltage or current or fluid flow when subjected in sequence to a series of processes such as amplification or transmission.

2. Algebraic Convolution and Generating Functions

When convolution is performed between two or more groups of 'things', and the 'things' can be categorized numerically, we can make use of some rules of algebra as will be shown presently. When, on the other hand, the categorisation of the 'things' is not numerical, we can do no more than head the columns of the necessary long-multiplication-sums-without-carrying in an appropriate manner such as that adopted in Tables 1.5, 1.6 and 1.7.

So far, we have been dealing with a non-numerical entity: colour! Let us now consider any entity which can be categorised numerically, or put in 'rank order', if the reader prefers the expression. This will enable us to express the operations of convolution in algebraic terms and, later, in terms of continuous functions enabling us to bring the calculus to our aid.

2.1 ADDITION OF RANDOM ERRORS IN GAINS OF TWO AMPLIFIERS IN TANDEM

Consider a telecommunications network containing only two linear amplifiers in tandem. I say 'linear' simply because I don't wish to have to consider non-linear distortion of the signal passing through them. Let us assume that these two amplifiers are supposed to have gains of 5 dB and 7 dB, respectively, and that unfortunately, they are not perfectly stable in the sense that the first (5 dB) amplifier's gain can drift up to ± 1 dB and that of the second can drift up to ± 2dB. Just for this particular example, let us assume that, for the 5 dB amplifier, there is an equal chance of finding, at any given moment, a gain anywhere between (5-1) dB and (5 + 1) dB or 4 dB and 6 dB and for the 7 dB amplifier an equally probable gain anywhere between 5 dB and 9 dB. The assumption that any gain figure within the given tolerances is equally likely to occur means that we have assumed rectangular statistical frequency distributions as shown in Fig. 2.1. Figures 2.1 (a) and (b) show the tolerances. Figures 2.1 (c) and (d) show quantised statistical distributions of gain, which assume that the gain can have only certain precise values. This is, of course, nonsense; but it is the first step towards constructing histograms, which we shall discuss presently. Note that since there are only three

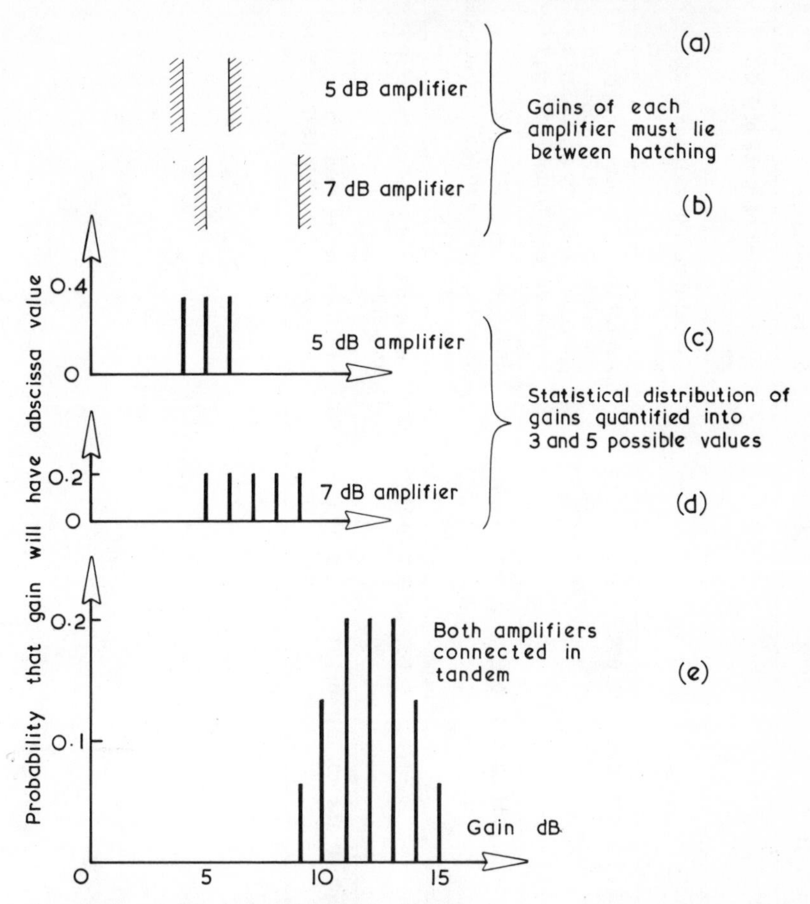

Fig. 2.1 Gain of amplifiers in tandem

possible gains allowed for the 5 dB amplifier and they are assumed to be equally probable, each one must have the value 1/3, because it is certain that the gain of this amplifier will be *either* 4 dB *or* 5 dB *or* 6 dB. Similarly, the gains of the 7 dB amplifier have probabilities each of 1/5 since there are five equal possibilities.

What overall gains are we likely to achieve for a signal that passes through each amplifier in turn? Table 2.1 shows the necessary convolution. For engineers unfamiliar with electronics it should be stated that decibels of gain for amplifiers in tandem add, because they are logarithmic units and the first amplifier will multiply the signal voltage (current) by a factor and the second by a further factor.

Figure 2.1 (e) shows the last line of Table 2.1 in graphical form.

Had there been more than two amplifiers we should have had to make repeated multiplications of the same kind as that shown in

Table 2.1 GAIN OF AMPLIFIERS IN TANDEM

4	5	6	7	8	9	10	11	12	13	14	15	Gains in decibels
	0.2	0.2	0.2	0.2	0.2							Probabilities of gains of 7 dB amplifier
0.33	0.33	0.33										Probabilities of gains of 5 dB amplifier
					0.067	0.067	0.067	0.067	0.067			Probabilities of occurrence of overall gain resulting from all possible gains of 7 dB amplifier with 4 dB gain from 5 dB amplifier
						0.067	0.067	0.067	0.067	0.067		Similarly: all gains of 7 dB amplifier with 5 dB gain from 5 dB amplifier
							0.067	0.067	0.067	0.067	0.067	Similarly: 6 dB gain from 5 dB amplifier
					0.067	0.134	0.201	0.201	0.201	0.134	0.067	Probabilities of having overall gains of 9, 10, 11, 12, 13, 14, 15 dB

Table 2.1. The process can be repeated without limit, but it rapidly becomes tedious.

Note:

(a) that the mean of the resulting statistical distribution, Fig. 2.1 (e), is the sum of the means of the constituent distributions, Figs. 2.1 (c) and (d), that is 12 dB = 5 dB + 7 dB.

(b) that the sum of the probabilities shown in the last line of Table 2.1 and in Fig. 2.1 (e) equals one.

(c) that the variance of the resulting distribution is the sum of the variances of the constituent distributions. The variance, or square of the standard deviation, is the second moment of the deviations from the mean. Thus a distribution $f(x)$ has a mean value $\bar{x}$ of the variate x:

$$\bar{x} = \frac{\sum_{i=-\infty}^{\infty} x_i f(x_i)}{\sum_{i=-\infty}^{\infty} f(x_i)} \tag{2.1}$$

and a variance:

$$\sigma^2 = \frac{\sum_{i=-\infty}^{\infty} (\bar{x} - x_i)^2 f(x_i)}{\sum_{i=-\infty}^{\infty} f(x_i)} \tag{2.2}$$

Thus, using Equation 2.2 and applying it to Fig. 2.1(c), then (d), then (e), we have remembering from Note (b) that

$$\sum_{i=-\infty}^{\infty} f(x_i) = 1,$$

$$2(1^2 \times 0.33) + 2(1^2 \times 0.2 + 2^2 \times 0.2)$$

$$= 2(1^2 \times 0.201 + 2^2 \times 0.134 + 3^2 \times 0.067)$$

or

$$2.66 = 2.68$$

which is near enough equality for our purposes. The variance, or the square-root of it, the standard deviation, is a measure of the spread or extent of the deviations from the mean. In communication engineering, the noise voltage that accompanies all signals can be regarded as statistical deviations of voltage around the mean value which is the signal itself. Thus, the variance of these deviations of voltage around the mean voltage is proportional to the noise power, the coefficient of proportionality being the reciprocal of the electrical resistance across which the noise voltage appears.

Table 2.1 would have been rather more simple if we had known about Note (a), because the top line of the table which contained the values of gain to be used as column headings for the convolution could have been restricted only to the following gain values: −3, −2, −1, 0, 1, 2, 3 dB, namely the variations around each of the two means, 5 and 7 dB. After concluding the convolution we would then have scaled the final axis of abscissae by adding $5 + 7 = 12$ dB to all the values.

Now, in Fig. 2.1 (c), (d) and (e) we have used single verticals or ordinates to describe the statistical distributions of gain, but it is more usual to use vertical blocks having finite widths such that the whole *area* of the distribution is accounted for. This makes no difference to the arithmetic convolution, but it helps us to slide painlessly from discontinuous histograms to continuous distributions where all gains between the given tolerances are taken into account instead of only those indicated by the presence of a vertical marker. Thus, Fig. 2.2 tells us that the probability of finding an overall gain anywhere between, for example, 9.5 dB and 10.5 dB is 0.134.

Let us refer back to Table 2.1 and consider how we came to choose the particular columns into which the probabilities resulting from the various multiplications should go. The fourth line of the table uses columns headed from 9 to 13 dB, and we said to ourselves 'if the probability that the 7 dB amplifier has a gain of 5 dB is 0.2 and the 5 dB amplifier has a gain of 4 dB with a probability of 0.33, then the probability of having a gain of $5 + 4 = 9$ dB is $0.2 \times 0.33 = 0.067$'. We then repeated the argument for the same 4 dB gain for the 5 dB amplifier, taking the other gains admitted for the 7 dB amplifier, thus filling in (for that line of the convolution) the columns 10, 11, 12, 13 dB. In other words, we found that whilst the probabilities were multiplied, the column headings were added. How can we express this in algebraic notation? Question: when two quantities are multiplied together have they any attribute that adds? Answer: yes, their powers or indices; thus

$$a^m \mathrm{x} a^n = a^{m+n} \tag{2.3}$$

but, of course,

$$a^m \mathrm{x} b^n \neq (ab)^{m+n} \tag{2.4}$$

so, if we want to multiply together two probabilities $\mathrm{f}(x)$ and $\mathrm{h}(y)$, for example, we cannot use the index or power method applied directly to the probabilities themselves, because usually $\mathrm{f}(x) \neq \mathrm{h}(y)$ and, furthermore, although we want to multiply a probability by another probability, we don't usually wish to raise a probability to a power except in cases where the two probabilities in question are equal. Suppose, however, that we use a column indicator or marking factor u^n such that a probability attached to the nth value of the variate is multiplied by u^n. Thus, for convenience of convolution we should multiply $\mathrm{f}(x_1)$ by u^1 and $\mathrm{h}(y_3)$ by u^3 and, in fact, any function of the variate values x_n and y_n such as

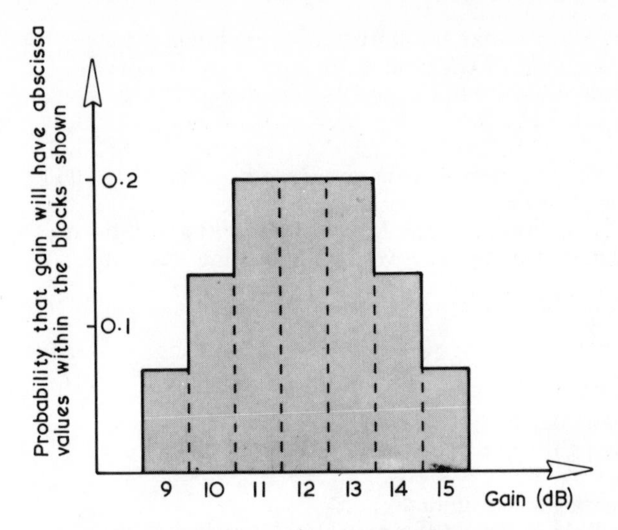

Fig. 2. 2 Histogram of gain of amplifiers in tandem

$f(x_n)$ or $h(y_n)$ would be multiplied by u^n. Do not worry about what value we are to allocate to u, because it is nothing more than a marking factor telling us into which column of a convolution a given probability product shall go. After the convolution has been accomplished we can let $u = 1$ when all the factors in powers of u will also be one, and u, having rendered us useful service, can be made to disappear. There is a subtlety here which must be understood. Although x and y are two separate variates, the multiplication of both $f(x_n)$ and $h(y_n)$ by u^n can take place only when the numerical values of x_n and y_n are equal. If we revert to our amplifier-gain problem set out in Table 2.1, we can multiply the probability of occurrence of a gain of 5 dB in the one amplifier by u^5, but we can multiply *only* the probability of occurrence of *5 dB* of gain in the other amplifier by u^5. If we multiplied by u^5 the probability of occurrence of 7 dB (for example) gain in the second amplifier, we should get into terrible confusion.

The convolution shown in Table 2.1 can now be indicated symbolically as

$$(0.2u^5 + 0.2u^6 + 0.2u^7 + 0.2u^8 + 0.2u^9)\ (0.33u^4 + 0.33u^5 + 0.33u^6)$$

where the powers of u are the decibel figures and we can factor-out $0.2u^7$ from the first factor and $0.33u^5$ from the second to obtain

$$0.2 \times 0.33u^{12}\ (u^{-2} + u^{-1} + u^0 + u^1 + u^2)\ (u^{-1} + u^0 + u^1)$$

or $\quad 0.066u^{12}(u^{-3} + 2u^{-2} + 3u^{-1} + 3u^0 + 3u^1 + 2u^2 + u^3)$

which, after multiplication of the factor in parentheses by $0.066u^{12}$, will give the last line in Table 2.1 with each term in the correct column. The column markers u^n can be seen to have much use, but as already mentioned, after they have served their purpose we can put $u = 1$ and then the + signs take on their proper arithmetic function. For example, if we want to check that the sum of all the probabilities is one, we let $u = 1$ in the last expression above, thus getting

$$0.066(1 + 2 + 3 + 3 + 3 + 2 + 1) = 0.99$$

and the answer would have been precisely one if we had changed 0.066 into $0.06\dot{6}$, which it ought to have been in any case.

When a marking factor such as u is 'glued' to a statistical distribution, or any other function to be convolved with another, such as

$$f(x) = f(x_1) + f(x_2) + f(x_3) + f(x_4) + f(x_5) \tag{2.5}$$

so that

$$g(u) = f(x_1)u^1 + f(x_2)u^2 + f(x_3)u^3 + f(x_4)u^4 + f(x_5)u^5 \qquad (2.6)$$

or, sometimes more conveniently

$$g(u) = u^3[f(x_1)u^{-2} + f(x_2)u^{-1} + f(x_3)u^0 + f(x_4)u^1 + f(x_5)u^2] \qquad (2.6)$$

We call $g(u)$ the generating function of $f(x)$.

A rather nice example of use of the generating function is the following[1]:

What is the probability of getting a total of nine points in three throws of a die? The statistical distribution of probabilities of getting 1, 2, 3, 4, 5 or 6 in one throw of the die is

$$f(x) = 1/6 + 1/6 + 1/6 + 1/6 + 1/6 + 1/6$$

The generating function is, therefore,

$$g(u) = \frac{u^1}{6} + \frac{u^2}{6} + \frac{u^3}{6} + \frac{u^4}{6} + \frac{u^5}{6} + \frac{u^6}{6}$$

$$= 1/6\ (u^1 + u^2 + u^3 + u^4 + u^5 + u^6)$$

where the columns of the convolution are clearly the various numbers, from one to three times six, that result from those that are engraved on each face of the die and result from one throw of one die to three throws of the one die. Thus the powers of u must be 1, 2, 3, 4, 5 and 6. If we now throw the die three times, or three dice at once, we must convolve $g(u)$ with itself, thrice.

The term in u^9 will be the probability of getting a total of nine points out of three throws.

$$g^3(u) = 1/6^3(u^3 + 3u^4 + 6u^5 + 10u^6 + 15u^7 + 21u^8 + 25u^9 + 27u^{10}$$
$$+27u^{11} + 25u^{12} + 21u^{13} + 15u^{14} + 10u^{15} + 6u^{16} + 3u^{17} + u^{18})$$

Thus, the probability of getting a total of nine in three throws is, putting the u_s equal to one, $25/216 = 0.116$.

Note that

$$1/6^3(1 + 3 + 6 + 10 + 15 + 21 + 25 + 27 + 27 + 25$$
$$+ 21 + 15 + 10 + 6 + 3 + 1) = 1$$

that is, it is certain that a score of either 3 or 4 or 5 or 6 or ... or

17 or 18 will be achieved. The most likely scores are ten and eleven, with equal probabilities of $27/216 = 0.125$.

If the facets of the die had been marked 3, 5, 7, 9, 11, 13, then the indices of the marker u would have to be the same numbers, in order to find the probabilities for various obtainable scores. Note, also, that the numbers on the facets of the die do not need to be in arithmetic progression.

Convolution of rectangular distributions or functions is quite fun, because, having factored-out the factor that is common to each term of the distribution, such as $u/6$ in the expression for the generating function in our die-throwing example, all one has to do is write down rows of ones, being careful to get the right 'ones' in the right columns. Thus, for $g^3(u)$ we have Table 2.2. When the distribution is not rectangular the arithmetic becomes tedious, but is feasible up to five or so sequential convolutions. The dashed curve in Fig. 2.3 shows on arithmetic probability paper the cumulative probability resulting from the convolution of five unequal rectangular distributions. The cumulative probability results from the progressive addition of

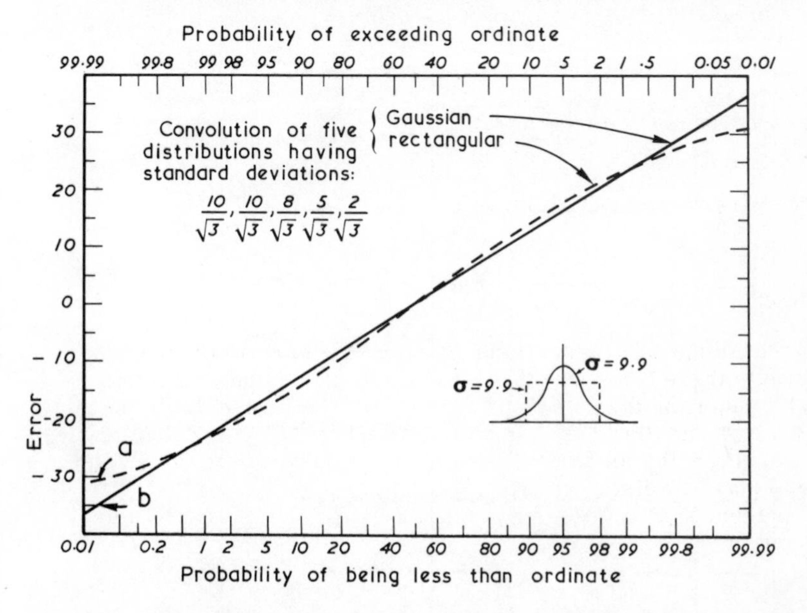

Fig. 2. 3 Cumulative probability of occurrence of overall errors from five devices in tandem (Courtesy of Royal Television Society)

each block of the histogram achieved by convolving the individual histograms representing each rectangular distribution. Thus a given ordinate on the cumulative histogram or distribution marks the probability that the variate (score from throwing five dice, for example) will not *exceed* the ordinate value.

Table 2.2 CONVOLUTION OF RECTANGULAR DISTRIBUTIONS

u^0	u^1	u^2	u^3	u^4	u^5	u^6	u^7	u^8	u^9	u^{10}	u^{11}	u^{12}	u^{13}	u^{14}	u^{15}	u^{16}	u^{17}	u^{18}
1	1	1	1	1	1	=	$6/u.\mathrm{g}(u)$											
1	1	1	1	1	1	=	$6/u.\mathrm{g}(u)$											
1	1	1	1	1	1													
	1	1	1	1	1	1												
		1	1	1	1	1	1											
			1	1	1	1	1	1										
				1	1	1	1	1	1									
					1	1	1	1	1	1								
1	2	3	4	5	6	5	4	3	2	1	=	$36/u^2.\mathrm{g}^2(u)$						
1	1	1	1	1	1	=	$6/u.\mathrm{g}(u)$											
1	2	3	4	5	6	5	4	3	2	1								
	1	2	3	4	5	6	5	4	3	2	1							
		1	2	3	4	5	6	5	4	3	2	1						
			1	2	3	4	5	6	5	4	3	2	1					
				1	2	3	4	5	6	5	4	3	2	1				
					1	2	3	4	5	6	5	4	3	2	1			
						1	2	3	4	5	6	5	4	3	2	1		
1	3	6	10	15	21	25	27	27	25	21	15	10	6	3	1	=	$216/u^3.\mathrm{g}^3(u)$	

The algebraic specification of a generating function of the type we have been using is

$$g(u) = \sum_{i=1}^{n} u^{x_i} f(x_i) \tag{2.7}$$

where x has n values, $x_1 x_2 x_3 \ldots x_n$ and $f(x)$ is the function describing the distribution or histogram or, for that matter, any function that we wish to convolve, either with itself or with other functions.

$f(x)$ can be and often is a polynomial in powers of x.

2.2 MOMENTS

Any curve plotted on a piece of paper has moments, like (i) the area between it and the axis of abscissae (the zeroth moment), or (ii) the mean value, or the centre of gravity (the first moment), or (iii) the moment of inertia, or the variance (the second moment) and so on. The nth moment Mn of a function $f(x)$ is defined as

$$M_n = \sum_{i=1}^{n} x_i^n f(x_i) \tag{2.8}$$

and therefore

$$M_0 = \sum_{i=1}^{n} f(x_i) \tag{2.9}$$

$$M_1 = \sum_{i=1}^{n} x_i f(x_i) \tag{2.10}$$

$$M_2 = \sum_{i=1}^{n} x_i^2 f(x_i) \tag{2.11}$$

and so on. Let us compare Equation 2.10 with Equation 2.1. First, we note that in Equation 2.1 the range of variation of i is from $-\infty$ to $+\infty$, whereas in Equation 2.10 it is only from 1 to n. There is really no difference here and I could have altered either equation to agree with the other. Equation 2.1 is really saying that the summation $\sum$ is to cover all possible values of i. Equation 2.10, on the other hand, says that since the function $f(x_i)$ is zero when i is either less than one or greater than n, it is a waste of time to pretend that values of i outside the range 1 to n have any useful significance. I shall not promise to use consistently either one of these two attitudes in what follows. The same argument applies to the limits indicated with definite integrals. Philosophically, all integrals should be between infinite limits and the integration process should start and stop at the values of the variable at which the function begins and ceases, respectively, to have a non-zero value. An exception to this is when, although the function has finite, non-zero values, we are not interested in examining it beyond the range given by the limits of summation or integration. In the latter case, putting non-infinite limits to the summation is a convenience which should not be allowed to conceal the truly general nature of integration or summation. One could even say that putting the function to zero, outside the range of interest is a legitimate thing to do, even though the initial function was not zero outside that range of interest.

A further point of difference between the two equations is that Equation 2.1 has a denominator (in fact, M_0), whereas Equation 2.10 does not. This difference is a matter of definition. My definition of the first moment (Equation 2.10) is not identical with the definition of the statistical mean. If we write

$$\bar{x} = \sum_{i=1}^{n} x_i \mathrm{f}(x_i) \; / M_0 \qquad (2.12)$$

$$= M_1/M_0$$

we can say that the mean is the normalised first moment. This discussion may seem to the reader to be unnecessary, but so often text books assume that

$$\sum_{i=1}^{n} \mathrm{f}(x_i) = 1$$

whence

$$M_0 = 1$$

and then

$$\bar{x} = M_1$$

without stating the fact explicitly that I have thought it better to clarify the matter.

Now let us look at Equation 2.2 and translate it into terms of moments:

$$\sigma^2 = \sum_{i=1}^{n} [\bar{x}^2 \mathrm{f}(x_i) + x_i^2 \mathrm{f}(x_i) - 2\bar{x}x_i \mathrm{f}(x_i)] \; / \sum_{i=1}^{n} \mathrm{f}(x_i)$$

$$= (\bar{x}^2 M_0 + M_2 - 2\bar{x}M_1)/M_0$$

$$= (\bar{x}^2 M_0 + M_2 - 2\bar{x}^2 M_0)/M_0$$

$$= M_2/M_0 - \bar{x}^2$$

or

$$\sigma^2 = M_2/M_0 - (M_1/M_0)^2 \qquad (2.13)$$

Equations 2.12 and 2.13 which express the mean on the one hand and the square of standard deviation on the other hand, or the variance, in terms of moments are only of real use if we can find an easy method of calculating moments, otherwise Equations 2.1 and 2.2 tell us how to calculate the mean and the variance in a straightforward, but tedious manner.

Consider the generating function defined by Equation 2.7

$$g(u) = \sum_{i=1}^{n} u^{x_i} f(x_i)$$

If we differentiate it with respect to u, we get

$$g'(u) = \sum_{i=1}^{n} x_i u^{x_i - 1} f(x_i) \tag{2.14}$$

A second differentiation yields

$$g''(u) = \sum_{i=1}^{n} x_i (x_i - 1) u^{x_i - 2} f(x_i) \tag{2.15}$$

If, in Equations 2.14 and 2.15 we put $u = 1$, we have

$$g'(1) = \sum_{i=1}^{n} x_i f(x_i)$$

$$= \bar{x} M_0 \text{ or } M_1 \tag{2.14a}$$

and

$$g''(1) = \sum_{i=1}^{n} x_i \; (x_i - 1) \; f(x_i)$$

$$= \sum_{i=1}^{n} x_i^2 \; f(x_i) - x_i f(x_i)$$

$$= M_2 - M_1 \tag{2.15a}$$

Thus, from Equation (2.14a) we have a method of calculating the mean of a function if we know its generating function and its zeroth moment, and from equations 2.13, 2.14a and 2.15a we can calculate the variance as

$$\sigma^2 = [g''(1) + g'(1)]/M_0 - [g'(1)/M_0]^2 \tag{2.16}$$

2.3 THE BINOMIAL DISTRIBUTION

Do we really ever come across a function whose generating function is easy to differentiate? Yes, and it is a much used function, too. It is the binomial distribution, which is a simpler case, but otherwise similar to our earlier problem of selecting coloured balls from an urn. Look back to Tables 1.3, 1.5, 1.6 and 1.7! Those tables showed the probabilities of selecting certain combinations of colours from two, three and four selections from an urn containing balls of three different colours. If we assume the urn contains only two types of ball, say black and white, it is then very easy to calculate the probabilities for various numbers of each type of ball from any number of selections, however large. If p is the proportion of white balls and $q = 1 - p$ the proportion of blacks in the urn and we make n selections, all we have to do is to raise $p + q$ to the nth power, thus

$$(p + q)^n = p^n + np^{n-1}q + \frac{n(n-1)}{2!} p^{n-2}q^2 + \frac{n(n-1)(n-2)}{3!} p^{n-3}q^3 + \ldots + \frac{n(n-1)\ldots(n+r+2)}{(r-1)!} p^{n-r+1}q^{r-1} \ldots + q^n \quad (2.17)$$

or, in terms of the notation of combinatory analysis,

$$(p + q)^n = C_n^0 p^n + C_n^1 p^{n-1}q + C_n^2 p^{n-2}q^2 + \ldots C_n^{r-1} p^{n-r+1}q^{r-1} + \ldots + C_n^n q^n \quad (2.18)$$

where the coefficient C_n^{r-1} of the rth term is the number of combinations of n things taken $r - 1$ at a time or

$$C_n^{r-1} = \frac{n!}{(r-1)!\,[n-(r-1)]!} \quad (2.19)$$

Note that

$$0! = 1! = 1 \quad (2.20)$$

The $(r + 1)$th term is perhaps easier to consider. It is

$$C_n^r p^{n-r} q^r = \frac{n!}{r!\,(n-r)!} p^{n-r} q^r \quad (2.21)$$

The probability that, out of n selections, we get all whites is $C_n^0 p^n$; $n - 1$ whites and one black: $C_n^1 p^{n-1}q$; $n - r$ whites and r blacks: $C_n^r p^{n-r} q^r$; all blacks: $C_n^n q^n$ and so on. I don't need to prove this, since it is a special case of the coloured balls (red, green and blue) example discussed earlier. Since $q = 1 - p$,

$$\begin{aligned}(p + q)^n &= 1^n \\ &= 1\end{aligned}$$

and the probability that one or other of the combinations of Equation 2.18 will arise in n selections is a certainty (addition law of probabilities). Let us consider a simple example. Suppose we make five selections from the urn of black and white balls, or toss a coin five times, or toss five coins at once. We have $n = 5$ and

$$\begin{aligned}(p + q)^5 &= C_5^0 p^5 + C_5^1 p^4 q + C_5^2 p^3 q^2 + C_5^3 p^2 q^3 + C_5^4 p q^4 + C_5^5 q^5 \\ &= p^5 + 5p^4 q + 10p^3 q^2 + 10p^2 q^3 + 5qp^4 + q^5\end{aligned} \quad (2.22)$$

Note the symmetry of the coefficients. This is because

$$C_n^r = C_n^{n-r}$$

Figure 2.4 shows Equation 2.22 when $p = 1/2$, $p = 1/3$ and $p = 1/10$. Now these three histograms each have a mean value of the variate; the latter being the number of white balls selected out of a total of five

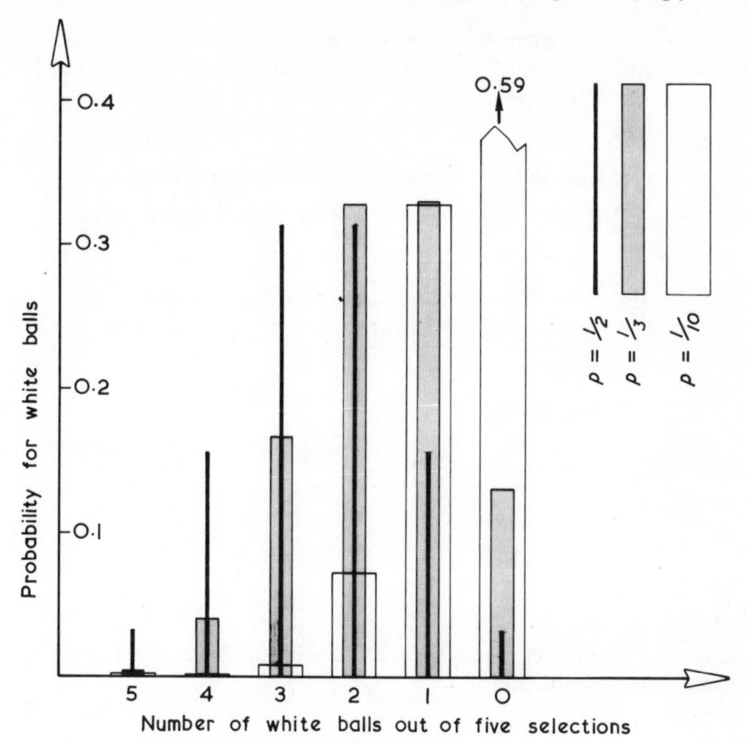

Fig. 2.4 Binomial theorem

trials or selections. It is necessary to understand what the 'mean number of white balls selected in five trials' really means. Suppose we say that (to take the case where $p = 1/3$ as an example) a five-selection trial resulting in five white balls will occur for 0.4% of the vast number of trials, because the probability for five whites is shown by the $p = 1/3$rd histogram to be about 0.004. A five-selection trial resulting in four whites (and a black) will occur for 4% of the trials; three whites for 17% of trials, two whites and one white will each account for about 33% of the trials and finally no whites will occur for about 13% of the trials. So, out of a vast number of five-selection trials, the average number of whites per trial will be (using Equation 2.1 and noting that $\sum_{i=1}^{5} f(x) = 100\%$)

$$(1/100)\ (0.4 \times 5 + 4 \times 4 + 17 \times 3 + 33 \times 2 + 33 \times 1 + 13 \times 0) = 1.68 \tag{2.23}$$

Having calculated the mean number of white balls per trial from Equation 2.1, let us now do so by using the generating function. Our distribution function is Equation 2.22, but the basic description of the proportions of white and black balls in the urn is simply $p + q$, so if u^1 indicates a white ball and u^0 a black one, we have for the generating function of $p + q$, the binomial

$$g(u) = pu^1 + qu^0$$

where u^n means n whites and, of course, $u^0 = 1$ means no whites. So,

$$\begin{aligned} g^5(u) &= C_5^0 p^5u^5 + C_5^1 p^4qu^4 + C_5^2 p^3q^2u^3 + C_5^3 p^2q^3u^2 + C_5^4 pq^4u + C_5^5 q^5 \\ &= (pu + q)^5 \end{aligned} \tag{2.24}$$

In order to apply Equation 2.14a and evaluate the mean number of white balls $\bar{x}$, we differentiate $g^5(u)$ with respect to u and then put $u = 1$. Thus

$$\frac{d}{du}(pu + q)^5 = 5(pu + q)^4 p$$

whence, remembering that in this case $M_0 = 1$.

$$\bar{x} = 5(p + q)^4 p$$

and with $p = 1/3$ and $p + q = 1$

$$\begin{aligned} \bar{x} &= 5/3 \\ &= 1.\dot{6} \end{aligned}$$

This result is, of course, accurate, whereas Equation 2.23 was the result of slide-rule calculations.

Generally

$$\frac{d}{du}(pu + q)^n = n(pu + q)^{n-1} p$$

and with $u = 1$ and $p + q = 1$ we have

$$(d/du)g^n(u) = np \tag{2.25}$$

whence

$$\bar{x} = np \tag{2.26}$$

Differentiating a second time

$$\frac{d^2}{du^2}(pu + q)^n = n(n - 1)(pu + q)^{n-2} p^2$$

and again, with $u = 1$ and $p + q = 1$ we have

$$\frac{d^2}{du^2}(pu + q)^n = n(n - 1)p^2 \tag{2.27}$$

Putting Equations 2.25 and 2.27 into Equation 2.16 and remembering that $M_0 = 1$, we have

$$\begin{aligned} \sigma^2 &= n(n - 1)p^2 + np - n^2p^2 \\ &= np(1 - p) \\ &= npq \end{aligned} \tag{2.28}$$

Equations 2.26 and 2.28 for the mean and variance of a binomial distribution are of fundamental importance and are used as stepping-stones in the logical development of the Gaussian or Normal law for the distribution of random errors, using the binomial distribution as starting point.

One simple application of the binomial distribution is that employed by some workers in the field of 'subjective testing'. In, for example, television and radio broadcasting engineering, it is often necessary to collect together a group of people and ask them to watch for some impairment to the sound or television image so that the engineer can form an opinion as to how much impairment it would be both economic and feasible to allow in a broadcast programme. Often, each test observer is given a sheet of paper on which is written a scale of numbers to each of which is allocated a subjective criterion of picture (or sound) quality or of impairment. For example, Table 2.3 shows such a scale.

Table 2.3

Criterion number or score	Criterion of impairment
5	imperceptible
4	perceptible but not annoying
3	visible, slightly annoying
2	annoying
1	very annoying

When the test is completed, the papers are collected, the mean score is calculated and a binomial distribution having the same mean score is substituted for the actual results (e.g. Reference 2). There is no logical reason why a set of test results should take the shape of a binomial histogram, but the method is a convenient way of allocating a mathematically specified histogram to a set of results, particularly as the latter rarely forms a symmetrical histogram, (with the mean in the middle of the scale).

The reader will have noted that I have defined, in Equation 2.8 the nth moment of a function, but that I have not had occasion to make use of moments of higher order than the second. Although statisticians make use of moments in order to fit mathematical functions to actual data, I have not, in my engineering work, had occasion to do so and I shall not, therefore, insist upon taking the reader through that process. It can be found in many textbooks (e.g. Reference 3).

3. An Example of Algebraic Convolution

3.1 ADDITION (BY CONVOLUTION) OF RANDOM ERRORS OF SIX PARAMETERS IN THE PAL SYSTEM

In the PAL system of colour television there is a number of parameters of the coded colour signal whose errors or departures from the correct values lead to impairments in the colour television image on the viewer's receiver. In particular, the saturation of the colours (that is, the vividness or paleness) can be adversely affected by errors of three parameters that affect the ratio of the amplitude of the chrominance or colouring signal to the magnitude of the luminance or brightness signal. Furthermore, in the PAL system, errors of phase in the chrominance signal also affect the saturation of the colours, but indirectly, through their squares. To simplify this example, we will omit the various coefficients that, as multiplying factors, affect the actual results, but do not influence the general treatment of the problem.

Let us assume that both the gain errors and the phase-angle errors are randomly distributed (at any one time), but that their statistical distributions are rectangular in form. In passing along a television transmission network, the colour signal will acquire three types of gain error and three types of phase error, and the overall effect on saturation will be additive. We thus have to convolve three rectangular distributions of gain error with three distributions of squared phase-angle error, the original distributions of phase angles being, also, rectangular. We shall thus arrive at the overall statistical distribution of saturation errors, $F(x)$.

We can take the rectangular distributions of gain error as consisting of, say, seven ordinates each of height $1/7$ with a central one at zero gain, the two extremes at ± 1 unit of gain and the intermediate ones at $\pm 1/3$ and $\pm 2/3$ units of gain.

We shall return to the gain distributions, presently. In the meantime we must give thought as to how to deal with the phase-angle distributions.

What we have to do is to find out what kind of a statistical distribution results from the squaring of the abscissae values of a rectangular distribution. Let

$$f(x) = 1/2, \qquad -1 \leqslant x \leqslant 1$$

be the specification of the rectangular distribution and, for the moment, consider it to be continuous between $x = -1$ and $x = +1$. The value $\frac{1}{2}$ was chosen simply to ensure that the whole area under the rectangle was one, Fig. 3.1a. This explains why the ordinates in Fig. 3.1 are labelled 'probability density'. An actual value of probability is now represented by an area rather than by a single ordinate.

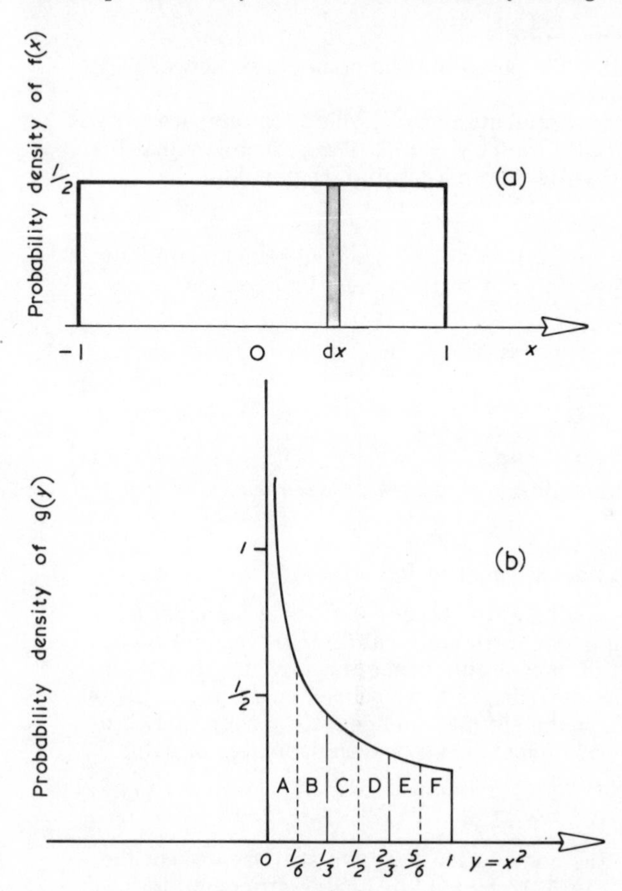

Fig. 3.1 Statistical distributions of phase angle and its square in a PAL colour television transmission

Moving in one bold step from discontinuous to continuous functions we can say that the probability that the phase, whose statistical distribution is shown as the rectangle in Fig. 3.1a, will lie within the hatched column of height $\frac{1}{2}$ and of width dx is

$$f(x)dx$$

This may be checked by integrating over the whole range of x, thus

$$\int_{-1}^{1} f(x)dx = \int_{-1}^{1} \tfrac{1}{2}dx$$

$$= \tfrac{1}{2} |x|_{-1}^{1}$$

$$= \tfrac{1}{2}(1 + 1)$$

$$= 1$$

That is, it is certain that the phase will be somewhere between −1 and +1.

Now what about the distribution $q(x^2)$, which for temporary convenience we shall call $q(y)$ where $y = x^2$? The probability that the squared phase angle will lie within a similar region dy is

$$q(y)dy$$

and if the phase angle lies within dx, the squared phase angle must lie within dy, hence the two probabilities must be equal or

$$f(x)dx = q(y)dy \qquad (3.1)$$

whence

$$q(y) = f(x)\frac{dx}{dy}$$

Since $y = x^2$ $x = y^{1/2}$ and $dx/dy = \tfrac{1}{2}y^{1/2}$ we have

$$q(y) = f(x)/2y^{1/2}$$

$$= \tfrac{1}{4}y^{1/2} \quad f(x) \text{ being equal to } \tfrac{1}{2}.$$

We can now plot $q(y)$ on Fig. 3.1b. Since $-1 \leqslant x \leqslant 1$ we must have $0 < y < 1$. This is a curious asymmetrical distribution and we have to convolve the result of convolution of the three rectangles of gain, $f(x)$, with three of these distributions of squared phase angle. We have already said that each rectangle may be adequately represented by the discontinuous function described by seven ordinates of value $\tfrac{1}{7}$ each and spaced as follows

$$x = -1, -\tfrac{2}{3}, -\tfrac{1}{3}, 0, \tfrac{1}{3}, \tfrac{2}{3}, 1$$

In order to make the convolution easy, we shall represent the distribution shown in Fig. 3.1b by four ordinates spaced as follows:

$$y = 0, \tfrac{1}{3}, \tfrac{2}{3}, 1$$

We must now decide what values these ordinates shall have. For this we divide the range of variation of $0 < y < 1$ into six regions A, B, C, D, E and F. The area of region A will be ascribed to the ordinate at $y = 0$. The area $B + C$ will be ascribed to the ordinate at $y = \tfrac{1}{3}$, the area $D + E$ to the ordinate at $y = \tfrac{2}{3}$ and finally, the area F to the ordinate at $y = 1$. Before ascribing areas to ordinates we shall normalise by dividing each area by the total area $A + B + C + D + E + F$.

Thus, we have the following calculations:

area under $q(y) = \frac{1}{4} \int_0^1 y^{-1/2} dy$

or

$$A + B + C + D + E + F = \left| y^{1/2}/2 \right|_0^1 = \frac{1}{2}$$

$$A = \frac{1}{4} \int_0^{1/6} y^{-1/2} dy = \sqrt{(1/6)}/2$$

Similarly

$$B + C = \{\sqrt{(1/2)} - \sqrt{(1/6)}\}/2$$

$$D + E = \{\sqrt{(5/6)} - \sqrt{(1/2)}\}/2$$

$$F = \{1 - \sqrt{(5/6)}\}/2$$

After normalising each area by dividing by ½ we have for each ordinate (Table 3.1)

Table 3.1

Ordinate value	Ordinate position
$\sqrt{1/6} = 0.408$	$y = 0$
$\sqrt{1/2} - \sqrt{1/6} = 0.299$	$y = 1/3$
$\sqrt{5/6} - \sqrt{1/2} = 0.206$	$y = 2/3$
$1 - \sqrt{5/6} = 0.087$	$y = 1$

So now we have the following convolution which, using the appropriate generating functions, may be written

$$G(u) = 1/7^3 (u^{-1} + u^{-2/3} + u^{-1/3} + u^0 + u^{1/3} + u^{2/3} + u^1)^3 \times (0.408u^0 + 0.299u^{1/3} + 0.206u^{2/3} + 0.087u^1)^3 \quad (3.2)$$

Note that the powers of the position or column marker u are precisely the values of the abscissae at which the samples or representative ordinates are placed. The above expression is slightly simpler if we factor 0.408^3 out of the second factor, thus getting

$$G(u) = 1.99 \times 10^{-4} (u^{-1} + u^{-2/3} + u^{-1/3} + u^0 + u^{1/3} + u^{2/3} + u^1)^3 (u^0 + 0.735u^{1/3} + 0.506u^{2/3} + 0.214u^1)^3$$

If we, temporarily and for convenience only, make the column spacings three times what they should really be, we can change the powers of u to integer or whole numbers, thus

$$G(u) = 1.99 \times 10^{-4} (u^{-3} + u^{-2} + u^{-1} + u^0 + u^1 + u^2 + u^3)^3 (u^0 + 0.735u^1 + 0.506u^2 + 0.214u^3)^3 \quad (3.2a)$$

and finally we have

$$G(u) = 1.01(0.0002u^{-9} + 0.001u^{-8} + 0.003u^{-7} + 0.007u^{-6}$$
$$+ 0.014u^{-5} + 0.023u^{-4} + 0.034u^{-3} + 0.048u^{-2} + 0.064u^{-1}$$
$$+ 0.078u^{0} + 0.09u^{1} + 0.097u^{2} + 0.1u^{3} + 0.096u^{4} + 0.087u^{5}$$
$$+ 0.075u^{6} + 0.06u^{7} + 0.045u^{8} + 0.033u^{9} + 0.021u^{10}$$
$$+ 0.013u^{11} + 0.006u^{12} + 0.003u^{13} + 0.0014u^{14} + 0.0004u^{15}$$
$$+ 0u^{16} + 0u^{17} + 0u^{18} \qquad (3.3)$$

The reader can check that if we put $u = 1$ the terms in the parentheses add to one. This is because the second factor in Equation 3.2 was normalised as indicated above Table 3.1 and the first factor was normalised by reason of the factor $1/7^3$. The factor 1.01 in front of equation 3.3 ought to have been 1.00 and is in error by 1% due to slide-rule calculations. Equation 3.3 is plotted in Fig. 3.2. The mean value is 2.93 in terms of the scale of abscissae, $3i$. The mean can be calculated directly from Equation 2.12 or from Equation 2.14a, noting that $M_0 = 1$. I calulated it by applying Equation 2.14a to the second factor in Equation 3.2. I knew that the mean of the convolution of the three rectangular distributions described by the first factor in Equation 3.2 would be zero, because the mean of each distribution is itself zero. I also knew that I need calculate the mean of only one of the three identical 'J' shaped (Fig. 3.b) distributions described by the second factor in Equation 3.2. Thus the overall mean of the abscissae in Fig. 3.2 will be three times

$$(d/du)[0.408(u^{0} + 0.735u^{1} + 0.506u^{2} + 0.214u^{3})] \text{ with } u = 1$$

which is the derivative, with respect to u, of the second factor in Equation 3.2a, but remembering that we must account for the factor 0.408 which is explicit in Equation 3.2, but is implicit in the factor 1.99 in Equation 3.2a. Thus

$$\overline{3i} = 3 \times 0.408(0.735 + 1.012 + 0.642)$$
$$= 2.93$$

The standard deviation can then be calculated, for example, by applying equation 2.16 and taking the square root.

First, we make use of Note (c) below Table 2.1; that is, the overall variance is the sum of the variances of the constituent distributions. Thus we have to calculate the variance of one of the three identical rectangular distributions, Fig. 3.1a, and add it to the variance of one of the three identical 'J' shaped distributions, Fig. 3.1b, and then multiply the sum by three and take the square root. Continuing to use the integer abscissae scale of $3i$ (rather than i), we have for the generating function of the rectangle

$$g_1(u) = 1/7(u^{-3} + u^{-2} + u^{-1} + u^{0} + u^{1} + u^{2} + u^{3})$$

and for the 'J' shaped distribution

$$g_2(u) = 0.408(u^0 + 0.735u^1 + 0.506u^2 + 0.214u^3)$$

$$g_1'(u) = {}^1\!/_7(-3u^{-4} - 2u^{-3} - u^{-2} + 0 + 1 + 2u + 3u^2)$$

Note that

$g_1'(1) = 0$; that is, the mean is zero.

$$g_1''(u) = {}^1\!/_7(12u^{-5} + 6u^{-4} + 2u^{-3} + 0 + 0 + 2 + 6u)$$

$$g_1''(1) = {}^{28}\!/_7$$

$g_2'(u) = 0.408(0.735 + 1.012u + 0.642u^2)$, as we already know,

and

$$g_2''(u) = 0.408(1.012 + 1.284u)$$

So

$$g_2'(1) = 0.975$$

and

$$g_2''(1) = 0.937$$

If σ_1^2 and σ_2^2 are the variances of one of the rectangles and one of the 'J' shaped distributions, respectively, we have from Equation 2.16

$$\sigma_1^2 = {}^{28}\!/_7 + 0 - 0^2$$
$$= 4$$

and

$$\sigma_2^2 = 0.937 + 0.975 - 0.975^2$$
$$= 0.962$$

So

$$\sigma^2 = 3(\sigma_1^2 + \sigma_2^2)$$
$$= 14.886$$

and

$$\sigma = 3.85$$

3.2 THE CENTRAL LIMIT THEOREM

Let us discuss Fig. 3.2 for a moment. First, the scale of abscissae. When x or y were one, we had $i = 1$, so we can re-introduce the original variates x and y into Fig. 3.2 merely by dividing the $3i$ scale by three. Equation 3.3 shows that the overall statistical distribution extends from $3i = -9$ to $3i = +18$ or $-3 < i < 6$. Thus the convolution of three rectangles each with a range of ±1 with three squared-variate distributions (the 'J' shaped ones), each of whose variate ranges are 0 to 1 yields a range going from $3(-1) = -3$ to $6 \times 1 = 6$ which is, or course, obvious. What is perhaps not so obvious are the very small prob-

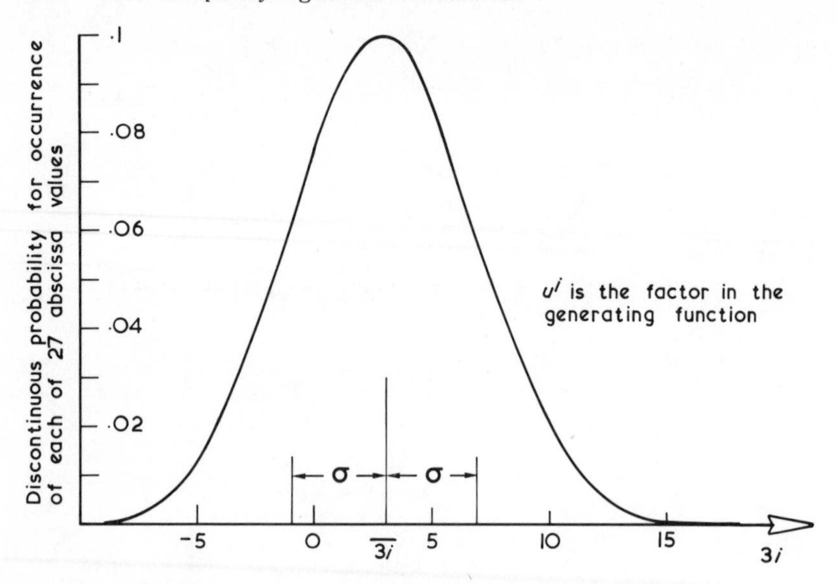

Fig. 3. 2 Overall statistical distribution of saturation errors in a PAL colour television transmission

abilities of occurrence of values at and near the extremes of the range of the variate. The probability of finding an overall value of saturation error of −3 is 0. 0002 and for an error of +5 it is 0. 0004 (see Equation 3. 3 or Fig. 3. 2). It is perhaps, surprising that such a smooth symmetrical (not perfectly symmetrical, of course) curve (Fig. 3. 2) results from convolving such abruptly terminating figures as rectangles with such asymmetrical figures as the 'J' shaped curves. If two equal rectangles are convolved, the result is a triangle, but the more convolutions are involved the smoother and the more symmetrical the result becomes. This phenomenon demonstrates (but does not prove) the validity of the 'Central Limit Theorem', which I shall not attempt to prove (too difficult!), (the gist of a proof is given on pages 16-20 of Reference 4) but which can be stated in the following form:

Consider a number of statistical distributions or functions in general

$$f_1(x_1), f_2(x_2) \ldots f_i(x_i)$$

having variates $x_1 \ldots x_i$ that are expressed in terms of the same unit.

Then, first, if

$$X = x_1 + x_2 + \ldots + x_i$$

and

$$F(X) = f_1(x_1) * f_2(x_2) * \ldots * f_i(x_i)$$

where * means 'convolved with',

then secondly, $F(X)$ will become more and more like the Gaussian or Normal law of errors as the number of functions $f_i(x_i)$ increases. When $i \to \infty$,

$$F(X) \to 1/\{\sigma\surd(2\pi)\}\exp(-X^2/2\sigma^2) + M_1 \text{ where } M_1 = \sum_{i=1}^{i\to\infty} \bar{x}_i$$

which is the Gaussian or Normal law much used in statistics and electronic engineering in connection with problems of electrical noise and signal-to-noise ratio. I have had to add M_1, the overall mean, or mean of means, to the Gaussian law, because the latter is so designed as to have a mean value of zero, whereas $F(X)$ may well have a non-zero mean. A proviso which can be important, but which is not so in the problems we have mentioned, is that the variabilities or standard deviations of the functions $f_1, f_2 \ldots\ldots f_i$ must not differ too widely. The expression

$$1/\{\sigma\surd(2\pi)\} \exp -X^2/2\sigma^2$$

can be derived directly from Equation 2.21 when n becomes very large and r is so arranged that the $(r + 1)$th term of the binomial expansion is, by a translation of the scale of abscissae, made to coincide with the term of maximum value in the whole expansion.

4. Mathematical Convolution

4.1 INTRODUCTION

The reader should, by now, possess a certain glow of self-confidence and feel that, at least for statistical applications, he could put algebraic or arithmetic convolution to good use, whether the problem involves electric-signal errors which have to be combined in order to obtain a statistical picture of an overall situation, or the chemical state of a fluid in some refining process, or even the addition or cumulative effects of tolerances on parts of a crankshaft in motor car engineering works.

We must now give thought to the convolution of continuous functions rather than the quantised algebraic functions such as those shown in Fig. 2.1(c), (d) and (e) or that expressed by Equation 2.5. In electronic engineering, particularly, convolution of continuous functions is of the greatest importance and could be used to the complete exclusion of Fourier analysis, although such an exclusion would discard most of the very useful and revealing way of thinking that has come to us from the telephone engineers. The concept of channels in a frequency spectrum such as those allocated to sound and television broadcasting transmissions or those allocated to voice signals in a frequency-division-multiplex telephone transmission could be replaced by the concept of sets of time-functions, ($\sin \omega t/\omega t$), which would just fill each frequency channel with signal power. Such a way of thinking would certainly increase the complexity of the task of the International Frequency Registration Board which would have to change its name to something like the International Non-interfering Time-Function Registration Board!

The electrical and electronics engineers will know that it is possible to calculate the transfer function of a cascade of tandem-connected linear circuits, provided they are linked with isolating devices such as transistors or valves, by multiplying together the individual transfer functions.

Thus, in Fig. 4.1, if

$V_2/V_1 = \mathrm{f}_1(j\omega)$

$V_3/V_2 = \mathrm{f}_2(j\omega)$

$V_4/V_3 = \mathrm{f}_3(j\omega)$

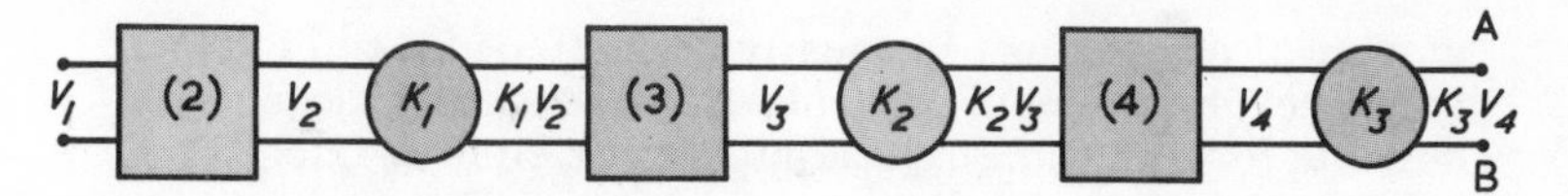

Fig. 4.1 A cascade of electric circuits
The attenuation or amplification factors K represent devices which connect circuit (2) with circuit (3), circuit (3) with circuit (4) and V_4 to the actual output terminals A and but only in the left-to-right direction in the figure; that is, they protect each circuit to their left from being influenced by the circuit to their right.

where ω is the angular frequency of a sinusoidal excitation $v_1 = V_1 \sin \omega t$ and, as usual in electrical engineering, $j^2 = -1$, and the functions f_1, f_2 and f_3 being complex and K_1, K_2 and K_3 being real constants (of amplification, for example), we may write, from consideration of the theory of alternatingcurrent,

$$V_4/V_1 = K_1K_2K_3f_1(j\omega)f_2(j\omega)f_3(j\omega)$$

The function V_4/V_1 is itself a function of angular frequency ω, and the expression for it tells us the ratio of the amplitude of v_4 to that of v_1 and the phase angle of v_4 after passage through the whole network. The ratio of amplitudes may be obtained by finding the modulus of the complex quantity V_4/V_1 and the phase by finding the argument of it. The whole process, however, is restricted to the hypothesis that the excitation is sinusoidal. Thus V_4/V_1 as a function of ω will tell us the effect of the cascade of circuits upon a sinusoidal excitation at angular frequency ω. It is usual to plot a 'frequency/response' curve which will, at a glance, tell the amplification or attenuation imposed on sinusoidal excitations at various angular frequencies. Similarly, a 'phase/frequency' curve may be plotted. Well and good! To the electrical engineer, at least, relatively simple and closely allied to the physical phenomena involved, provided the excitation is sinusoidal! Suppose it isn't! what then? A vast literature, involving some pretty heavy mathematics has risen to set out methods of solving the problem for non-sinusoidal excitations. The first method is, or course, to write out and solve the differential equations for the currents or voltages or electric charges to be found in the circuits under consideration. The second method which achieved widespread use was to employ Heaviside's operational calculus. The third method, in much use still, is to use infinite integrals based on that of Fourier. The Laplace transform and the Carson integral are the two favourites, with the former being a good deal more popular than the latter. The method involving these infinite integrals, or more correctly, integrals between infinite limits, is as follows:

(a) Find the response of the circuit being studied to sinusoidal excitations at all frequencies

(b) Analyse the excitation function (the excitation in the case of electrical engineering is usually a function of time) into a frequency spectrum of sinusoidal excitations by means of a suitable form of the Fourier integral or similar integrals such as the Laplace integral or the Carson integral

(c) Multiply the response as a function of frequency from (a) by the spectrum from (b)

(d) Find, by the other form of the Fourier integral, called the Mellin inversion integral, that function of time which would produce the spectrum found by (c)

The foregoing operations are by no means simple or easy to understand, but since convolution inevitably comes into some of the proofs required to substantiate the methods, it is hoped that a thorough understanding of it will help the student to plough his way through the morass of mathematics with a firm grip on what is really going on rather than sole reliance on algebraic logic. Furthermore, some problems are more tractable by direct use of convolution rather than plodding through operations (b), (c) and (d).

4.2 CONVOLUTION OF SMOOTH FUNCTIONS

Let us revert to Fig. 2.1 and Table 2.1 in which we convolved arithmetically the decibel gains of two amplifiers connected in tandem. Let us express the statistical variations of the gain of the 5 dB amplifiers, Fig. 2.1(c), and those of the 7 dB amplifier, Fig. 2.1(d), respectively, as

$$\mathrm{f}(x) \text{ and } \mathrm{g}(y)$$

We now know that the probability of finding an overall gain of

$$u = x + y \tag{4.1}$$

when x and y have the special values x_1 and y_1 is

$$\mathrm{f}(x_1)\mathrm{g}(y_1)$$

That is, $\mathrm{f}(x_1)$ being the probability of finding a gain x_1 in the 5 dB amplifier and $\mathrm{g}(y_1)$ being the probability of finding a gain y_1 in the 7 dB amplifier, fg is the probability of finding both of these gains simultaneously. Table 2.1 shows, in its last line, the resulting statistical distribution and so does Fig. 2.1(e). But the value of $u_1 = x_1 + y_1$ might also arise from several different pairs of values of x and y; thus, Table 2.1 shows that an overall gain of, for example, 11 dB can arise in three different ways, namely

Table 4.1

Overall gain		Individual gains
4 dB + 7 dB	or	$x_1 = 4\ y_1 = 7$
5 dB + 6 dB	or	$x_1 = 5\ y_1 = 6$
6 dB + 5 dB	or	$x_1 = 6\ y_1 = 5$

and it is the overall gain, u, that we must make explicit. Now, from Equation 4.1 we have

$$y = u - x$$

so we can write for the probability of finding a gain

$$u = x + y$$

$$\mathrm{P}(u) = \sum_{x=4}^{6} \mathrm{f}(x)\mathrm{g}(u - x) \qquad (4.2)$$

or

$$\mathrm{P}(u) = \sum_{y=5}^{9} \mathrm{f}(u - y)\mathrm{g}(y) \qquad (4.3)$$

The summation sign tells us to add up all the various ways that there are of obtaining each value of u; thus, taking Equation 4.2 as an example, we have Table 4.2.

Table 4.2 THE SEVERAL WAYS BY WHICH OVERALL GAIN OF TWO AMPLIFIERS IN TANDEM CAN ARISE

When u is	P(u) is
9	f(4)g(5) + f(5)0 + f(6)0
10	f(4)g(6) + f(5)g(5) + f(6)0
11	f(4)g(7) + f(5)g(6) + f(6)g(5)
12	f(4)g(8) + f(5)g(7) + f(6)g(6)
13	f(4)g(9) + f(5)g(8) + f(6)g(7)
14	f(4)0 + f(5)g(9) + f(6)g(8)
15	f(4)0 + f(5)0 + f(6)g(9)

The zeros represent the function g(y) for values outside its range of $5 \leqslant y \leqslant 9$.

Equations 4.2 and 4.3 may be written symbolically, using the asterisk notation for convolution,

$$P = f * g \tag{4.4}$$

and it is obvious from our previous work that the commutative property applies

$$P = g * f \tag{4.5}$$

Since

$$P(u) = P(x + y) \tag{4.6}$$
$$= P(y + x)$$

convolution allows the associative property of ordinary addition to apply to the variables involved.

We shall soon be introducing the delta function or unit impulse, so the following remarks will be helpful later on. Consider two functions to be convolved. As before, let the variable be restricted to integer values rather than be continuously variable. Let the functions be $f(x_i)$ and $g(y_i)$ as shown in Fig. 4.2. $f(x_i)$ was quite arbitrarily chosen, but $g(y_i)$ was purposely restricted to only one ordinate of

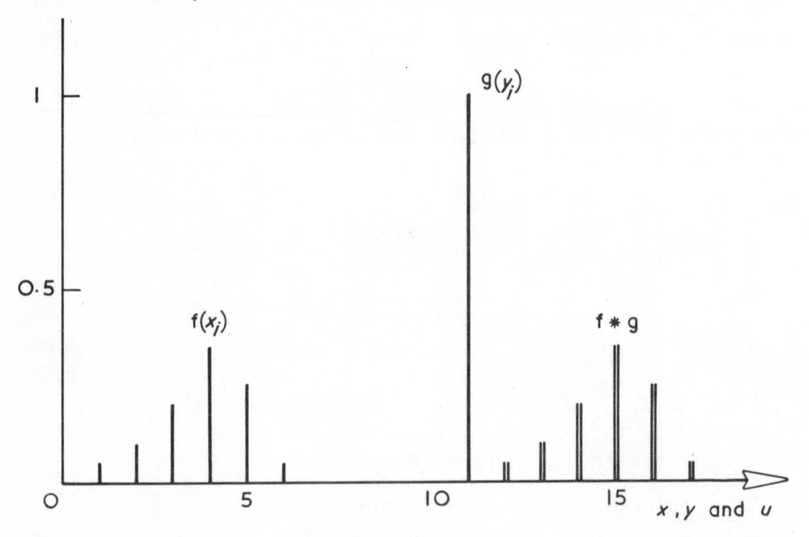

Fig. 4.2 Effect produced by convolving any function by the delta function

height 'one' at abscissa value $y = 11$. All that convolving f by g = 1 (at a single value of abscissa) has done is to move the function f along the axis of abscissae by an amount $y = 11$; the function f is otherwise unchanged. Thus a function, unity, at z along the x-axis (or u-axis) is an 'operator' which when convolved with another function f shifts it by an amount z along the x-axis. This will be proved for continuous functions later. Since the single-valued 'operator' g had unit value all it did to the ordinates of $f(x_i)$ was to translate them horizontally and multiply each one of them by one. Its additive effect on the abscissae results from the same property of convolution as gave rise to Note (a) below Table 2. 1; namely, that the mean of a convolution is the sum of the means of the constituent functions.

Now let us try to convert the convolution product in equation 4. 2 which is between functions whose variables move in steps, $x = 4, 5, 6$, for example, into a convolution of continuous functions. The reader may remember that in dealing with the example in Chapter 3 we had to convert a smooth or continuous function into a series of ordinates, Fig. 3. 1(b). What we did was to ascribe the value of an area such as $B + C$ to the ordinate going through the middle of it (except in the cases of the extreme ordinates at $y = 0$ and $y = 1$ where we took the area to one side only, since there was no area on the other side). We shall now use the same idea, only the other way round. Thus, consider an ordinate of the product

$$f(x_i)g(u - x_i)$$

in the summation from Equation 4. 2

$$P(u) = \sum_{l=1}^{3} f(x_i)g(u - x_i), \text{ where } x_i = i + 3 \qquad (4.7)$$

This ordinate may be regarded, for the moment, as a function of x_i only, since it is the actual operation of summation ($\sum$) with which we are concerned rather than with the final answer which is, of course, a function only of u. We can regard u, for the moment, as a constant parameter. We may thus ascribe to the ordinate $f(x_i)g(u - x_i)$ at x_i, the area $f(x)g(u - x)dx$, whence Equation 4. 7 which tells us to add all such ordinates, now tells us to add all such areas, thus

$$P(u) = \int f(x)g(u - x)dx \qquad (4.8)$$

The reader may well ask if the f and the g in Equation 4. 8 are the same as the f and the g in Equation 4. 2. The answer is, strictly, 'no', but it is reasonable to say that the f and g of Equation 4. 8 represent smooth functions that will have the same shape as the envelope of the ordinate values of the f and g of Equation 4. 2. The f and g of Equation 4. 8 are any continuous function in the class of those which could be obtained by passing the f and the g from Equation 4. 2 through an appropriate low-pass filter, either electrically or mathematically. The cut-off frequency of the filter would have to be at least $\frac{1}{2}(x_i - x_{i-1})$ or $\frac{1}{2}\Delta x$ where Δx is the spacing between the ordinates of $f(x_i)$ and of $g(x_i)$. We cannot prove this, because at present we haven't the knowledge to do so and in any case such matters are outside the scope of this book. The above remarks rely on the functions f and g being dimensionless numbers such as probabilities or gains of amplifiers, but even in these circumstances, in transferring from a set of discontinuous ordinates to a continuous curve, the ordinate dimension must change from its discontinuous value, say probability, to a probability density so arranged that the total area under the curve becomes unity, just as in the case of the discontinuous ordinates the total sum of their values must be unity. Figure 3.1 illustrates this point.

Note that I have omitted limits of integration from Equation 4.8. This matter requires some discussion. We want the convolution integral to be a function of u, not x and although the integral is often given in textbooks between infinite limits, that is

$$\int_{-\infty}^{\infty} f(x)g(u-x)dx$$

this is merely a method of saying that all possible products of ordinates of f and g must be taken on the assumption that f and g are continuous functions in the ranges $-\infty < x < \infty$ and $-\infty < y < \infty$. In fact, the integral will often be written

$$\int_{-\infty}^{u} f(x)g(u-x)dx \tag{4.9}$$

and this is fine for functions f and g that are discontinuous in the mathematical sense. For such functions, however, careful thought is required at each stage of the calculation, as the following example will show. It is not an easy one and after the reader has gone through it carefully his understanding of convolution will be better than that of most engineers and scientists.

4.2.1 An example

Figure 4.3 shows two discontinuous, but otherwise smooth functions:

$$f(x) = \sin(x - \pi/10),\ \pi/10 < x < 11\pi/10 \tag{4.10}$$

and

$$g(y) = (1/2)\sin(y/2 - 2\pi/3),\ 4\pi/3 < y < 10\pi/3 \tag{4.11}$$

Required

$$P(u) = \int_{-\infty}^{u} f(x)g(u-x)dx$$

First we make use of Note (a) below Table 2.1 in Chapter 2; namely, that the mean of the function $P(u)$ is the sum of the mean of $f(x)$ and of $g(y)$.

$f(x)$ and $g(y)$ happen to have even symmetry around their mean values, which occur at their maxima. This is obvious. Thus, the means $\bar{x}$ and $\bar{y}$ occur when each of the two functions reach their maxima. Hence

$$\sin(\bar{x} - \pi/10) = 1$$

and

$$(\tfrac{1}{2}) \sin (\bar{y}/2 - 2\pi/3) = \tfrac{1}{2}$$

whence

$$\bar{x} = 1.885 \text{ and } \bar{y} = 7.34$$

Thus, the mean of the final convulution product $P(u)$ will be at

$$\bar{u} = \bar{x} + \bar{y}$$

$$= 9.225$$

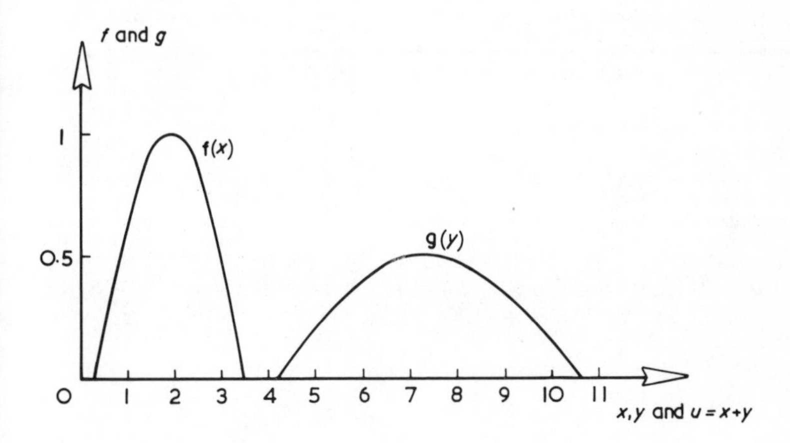

Fig. 4. 3 Convolution of two half-sine waves

We can now simplify our thinking by shifting $f(x)$ and $g(y)$ shown in Fig. 4. 3 to the position shown in Fig. 4. 4. When the convolution is completed we must remember to add the quantity $\bar{u} = 9.225$ to the mean of the result. In fact, the mean of the convolution of the two functions shown in Fig. 4. 4 will be zero so the final mean will be $\bar{u} = 9.225$. The two functions F and G shown in Fig. 4. 4 are now

$$F(x) = \cos x$$

and

$$G(y) = (\tfrac{1}{2}) \cos (y/2)$$

The process of convolution amounts to sliding G along the axis of abscissae as we make u vary and integrating the product of the two functions when they overlap. Thus if we leave F fixed and translate G as a function of u, we start the process with G slightly to the left of the position shown in Fig. 4. 5. Thus, if we take u, arbitrarily, as defining the mid-abscissa or mean value of the sliding function G, the starting value of u is

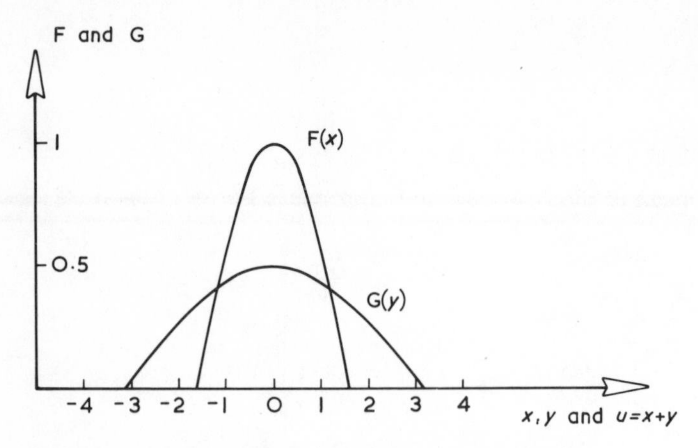

Fig. 4.4 Convolution of two half-sine waves

$$u_0 = -(\pi/2 + \pi)$$
$$= -3\pi/2$$

This is the value of u when the right-hand 'tail' of $G(u - x)$ just touches the left-hand 'tail' of $F(x)$, and the process of integration is valid (by symmetry), until $u = +3\pi/2$. Thus, Equation 4.9 becomes

$$P(u) = \int_{-\pi/2}^{u} (\cos x)\,(1/2) \cos \overline{(u - x/2)}\,dx \tag{4.12}$$

with

$$-3\pi/2 \leqslant u \leqslant 3\pi/2$$

But matter do not end here, because of the discontinuous nature of the two half-cosines F and G. Equation 4.12 remains valid only when

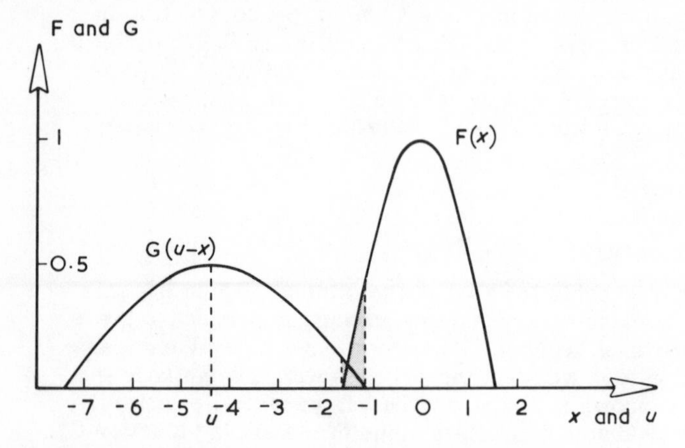

Fig. 4.5 Convolution of two half-sine waves

$-3\pi/2 \leqslant u \leqslant -\pi/2$. The starting value of $u, u_0 = -3\pi/2$, can be seen from Fig. 4. 5 to be slightly to the left of the position on the axis of abscissae marked 'u'. The finishing value, as far as Equation 4. 12 is concerned, is $u_1 = -\pi/2$, because when $u > -\pi/2$ the right-hand 'tail' of G exceeds $+\pi/2$, at which point $F = 0$ and we must stop integrating. Fig. 4. 6 shows $G(u_1 - x)$, that is, $G(u - x)$ at the final position for which Equation 4. 12 remains valid. Thus Equation 4. 12 now becomes

$$P_1(u) = \int_{-\pi/2}^{u+\pi} F(x)G(\overline{u - x}/2)dx \qquad (4.13)$$

with

$-3\pi/2 \leqslant u \leqslant -\pi/2$ that is, $u_0 \leqslant u \leqslant u_1$

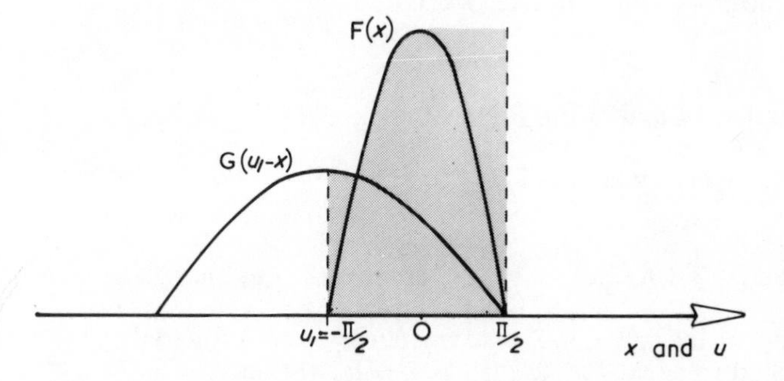

Fig. 4. 6 Convolution of two half-sine waves

What we are doing is to ensure, by suitable limitation of the range of variation of u, that the variable of integration, x, is always restricted to the region of overlap of the two functions F and G. Since the half-width of G is π and $u_0 = -3\pi/2$, we see that the upper limit of integration $u + \pi$ starts at $u_0 + \pi = -\pi/2$ and ends at $u_1 + \pi = \pi/2$.

We now consider what to do when the right-hand 'tail' of G exceeds $x = \pi/2$. Figure 4. 7 shows a position of G in this situation. It is plain to see that for this case the limits of integration are simply $-\pi/2 \leqslant x \leqslant +\pi/2$ and the starting value for u is $u_1 = -\pi/2$ and the finishing value occurs when the left-hand tail of G coincides with the right-hand tail of F, that is $u_3 = 3\pi/2, (\pi + \pi/2)$. However, reasons of symmetry show that we can stop at $u_2 = 0$, because the two functions, being themselves evenly symmetrical, will yield a convolution product with even symmetry around the abscissa value of zero. Thus our second and final version of equation 4. 12 is

$$P_2(u) = \int_{-\pi/2}^{\pi/2} F(x)G(\overline{u - x}/2)dx \qquad (4.14)$$

with

$-\pi/2 \leqslant u \leqslant 0$ that is, $u_1 \leqslant u \leqslant u_2$

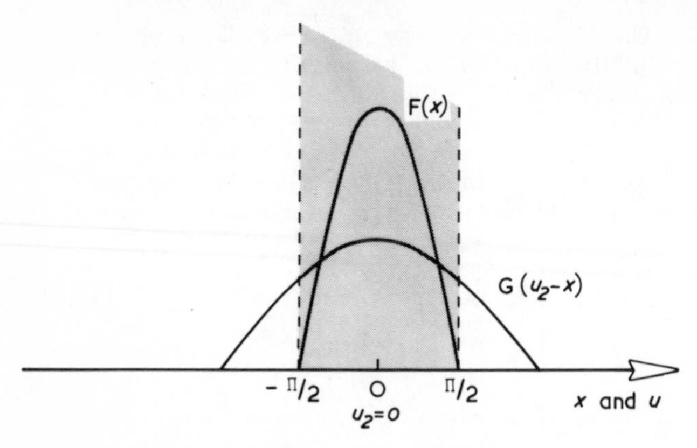

Fig. 4.7 Convolution of two half-sine waves

Equations 4.13 and 4.14 are of the form

$$I = \tfrac{1}{2} \int_a^b \cos x \cos (u/2 - x/2)dx$$

$$= \tfrac{1}{4} \int_a^b \cos (x/2 + u/2)dx + \tfrac{1}{4} \int_a^b \cos (3x/2 - u/2)dx$$

For the first integral let $x/2 + u/2 = z$, whence $x = 2z - u$ and $dx = 2dz$. For the second integral let $3x/2 - u/2 = w$, whence $x = 2w/3 + u/3$ and $dx = 2dw/3$. With these changes of variables we have

$$I = \tfrac{1}{2} \int_{u/2 + a/2}^{u/2 + b/2} \cos z dz + \tfrac{1}{6} \int_{3a/2 - u/2}^{3b/2 - u/2} \cos w dw$$

or

$$I = \tfrac{1}{2}[\sin (u/2 + b/2) - \sin (u/2 + a/2) + \tfrac{1}{3} \sin (3b/2 - u/2)$$

$$- \tfrac{1}{3} \sin (3a/2 - u/2) \qquad (4.15)$$

All we now have to do is to put the limits of integration from Equation 4.13 into Equation 4.15 and then the limits of integration from Equation 4.14 into Equation 4.15 to arrive at our two expressions for $P(u)$.

For Equation 4.13 we have

$$a = - \pi/2, b = u + \pi$$

whence, from Equation 4.15, we have

$$P_1(u) = \tfrac{1}{2} [\sin (u + \pi/2) - \sin u/2 - \pi/4) + \tfrac{1}{3} \sin (u + 3\pi/2)$$
$$+ \tfrac{1}{3} \sin (u/2 + 3\pi/4)]$$

or

$$P_1(u) = \tfrac{1}{3}[\cos u + \sqrt{2}\,(\cos u/2 - \sin u/2)] \tag{4.16}$$

with

$$-3\pi/2 \leqslant u \leqslant -\pi/2$$

For Equation 4.14 we have

$$a = -\pi/2 \quad b = \pi/2$$

and Equation 4.15 then becomes

$$P_2(u) = \tfrac{1}{2}[\sin(u/2 + \pi/4) - \sin(u/2 - \pi/4) + \tfrac{1}{3}\sin(3\pi/4 - u/2) + \tfrac{1}{3}\sin(3\pi/4 + u/2)]$$

or

$$P_2(u) = \frac{2\sqrt{2}}{3}\cos u/2 \tag{4.17}$$

with

$$-\pi/2 \leqslant u \leqslant 0$$

Figure 4.8 shows $P_1(u)$ with $-3\pi/2 \leqslant u \leqslant -\pi/2$ and $P_2(u)$ with $-\pi/2 < u \leqslant 0$, that is, $P(u)$. The portion of the curve for $u > 0$ was obtained by symmetry.

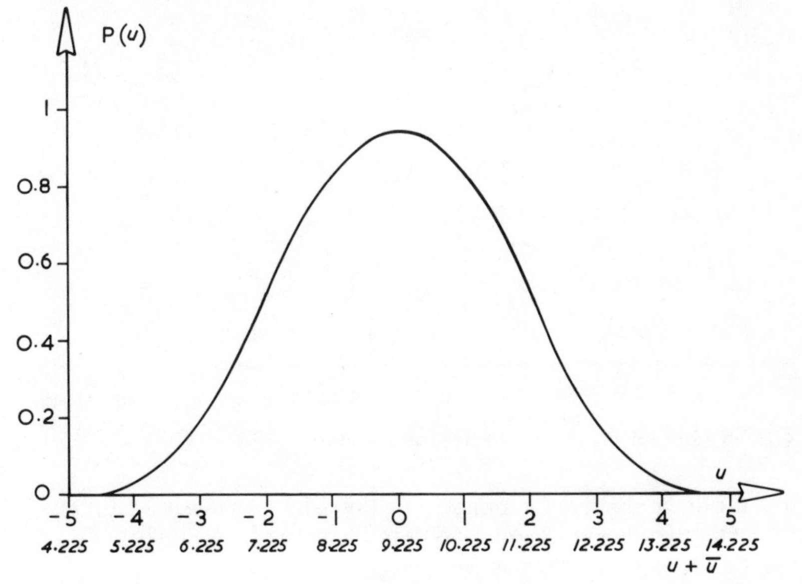

Fig. 4.8 Convolution of two half-sine waves

Well, I hope the reader found that easy, I didn't! At any rate, we can now say that the limits of integration must be carefully thought out and the upper limit may contain u when neither function is entirely overlapped by the other, but they can be given the limiting values of x when one function completely overlaps the other. Further, the permissible range of values of u must not allow the permissible range of values of x to be exceeded. Suppose that both functions are continuous within the range of $a \leqslant x \leqslant b$. In that case, the limits are a and b and if $-\infty < x < \infty$ and $-\infty < y < \infty$ then the convolution integral is taken between those limits. The only lack of generality in the example we have taken is that the functions F and G had even symmetry about their mean values, but this is not a serious limitation to a thorough understanding of the process.

4.2.2 Same example; alternative form of convolution integral

Nevertheless we shall now go through the whole problem again, but this time as an exercise, we shall choose the other form of the integral; that is, we shall make the narrower function F(x) the sliding one and the wider one G(y) the fixed function. With the knowledge we now have, we can immediately write down the convolution integral, Equation 4.9 as (see Fig. 4.9)

$$P_1(u) = \tfrac{1}{2} \int_{-\pi}^{u+\pi/2} \cos(u-y)\cos y/2\,dy \quad \text{with } -3\pi/2 \leqslant u \leqslant -\pi/2 \tag{4.18}$$

and

$$P_2(u) = \tfrac{1}{2} \int_{u-\pi/2}^{u+\pi/2} \cos(u-y)\cos y/2\,dy \quad \text{with } -\pi/2 < u \leqslant 0 \tag{4.19}$$

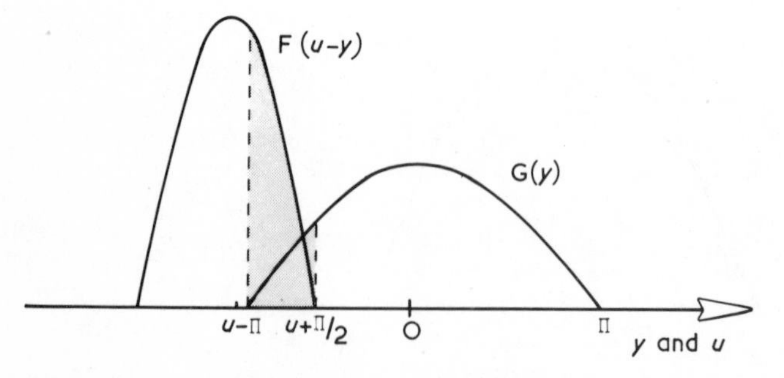

Fig. 4.9 Convolution of two half-sine waves

For $u > 0$ we use the even symmetry of P(u) and take the mirror image of the function from $-3\pi/2$ to 0 to obtain its shape from 0 to $3\pi/2$. Equations 4.18 and 4.19 are of the type

$$I = \tfrac{1}{2} \int_a^b \cos y/2 \cos(u-y)\,dy$$

$$= \tfrac{1}{4} \int_a^b \cos(u-y/2)\,dy + \tfrac{1}{4} \int_a^b \cos(u-3y/2)\,dy$$

If, in the first integral of the above equation, we let

$$u - y/2 = z$$

and in the second integral

$$u - 3y/2 = w$$

we have

$$I = -\tfrac{1}{2}\int_{u-a/2}^{u-b/2} \cos z\,dz - \tfrac{1}{6}\int_{u-3a/2}^{u-3b/2} \cos w\,dw$$

whence, finally we arrive at $P_1(u)$ and $P_2(u)$. The reader may care to check that the $P_1(u)$ of Equation 4.18 is equal to the $P_1(u)$ in Equation 4.16 and, similarly $P_2(u)$ in Equation 4.19 is the same as $P_2(u)$ in Equation 4.17. I have checked these!

4.2.3 A first example with infinite limits of integration

If we consider two functions f and g that are continuous over the whole range of x and y between $-\infty$ and $+\infty$, the limits of integration of the convolution take on these values. Let us, therefore, as an example, convolve the parabola $f(x) = x^2$ with the Gaussian error function or Normal curve of errors $g(y) = A \exp - a^2y^2$. Let us, first, normalise the area under $g(y)$ to unit value.

$$A\int_{-\infty}^{\infty} \exp(-a^2y^2)dy = 1$$

I shall not trouble the reader with the methods by which various integrations that we shall be considering are solved. Those readers who are familiar with advanced methods will not require such help and it is not the purpose of this book to teach branches of mathematics other than convolution. So, from any suitable table[5] of integrals

$$A\int_{-\infty}^{\infty} \exp(-a^2y^2)dy = 2A\int_{0}^{\infty} \exp(-a^2y^2)dy \text{ (by even symmetry of the integrand)} = A\sqrt{\pi}/a$$

whence

$$A = a/\sqrt{\pi}$$

so

$$g(y) = (a/\sqrt{\pi}) \exp(-a^2y^2)$$

Now let $u = x + y$ and our convolution integral becomes

$$P(u) = \int_{-\infty}^{\infty} f(x)g(u-x)dx \tag{4.20}$$

$$= \frac{a}{\sqrt{\pi}}\int_{-\infty}^{\infty} x^2 \exp\{-a^2(u-x)^2\}\,dx \tag{4.21}$$

If we let $u - x = z$ with $dx = -\,dz$, we have

$$P(u) = \frac{a}{\sqrt{\pi}} \int_{-\infty}^{\infty} (u - z)^2 \exp(-a^2z^2)dz \text{ (really: } -\int_{\infty}^{-\infty}$$

$$= a/\sqrt{\pi} \int_{-\infty}^{\infty} \{u^2 \exp(-a^2z^2) - 2uz \exp(-a^2z^2) + z^2 \exp(-a^2z^2)\} dz$$

The middle term integrates to zero due to its odd symmetry around $z = 0$ and the first and third terms, having even symmetry, allow ut to integrate from 0 to ∞ whilst doubling the answer. Thus

$$P(u) = 2a/\sqrt{\pi}\,[u^2 \int_0^{\infty} \exp - a^2z^2)dz + \int_0^{\infty} z^2 \exp(-a^2z^2)dz]$$

$$P(u) = u^2 + \tfrac{1}{2}a^2 \tag{4.22}$$

which is, perhaps, a rather surprising result. We find a parabola, similar to $f(x)$, but 'sitting' on a constant level of $\frac{1}{2}a^2$. If, whilst keeping the area under $g(y)$ equal to unity, we make it narrower and narrower, therefore taller and taller, by letting $a^2 \rightarrow \infty$, we find that $P(u) \rightarrow u^2$, the original parabola, but with abscissae $u = x + y$! This is in accordance with Fig. 4. 2, which showed that convolving a function with unit impulse or the delta-function merely translated the original function along the axis of abscissae by the abscissa value of the delta-function itself. Now

$$\frac{a}{\pi} \exp(-a^2y^2)$$

with $a \rightarrow \infty$ becomes the infinitesimally wide, infinitely tall function $\delta(y)$. There are many functions which for a certain parameter either tending to zero or to infinity become, in the limit, the idealised concept $\delta(y)$. For example, another form of the delta-function would be a rectangle of width a and of height $1/a$ with $a \rightarrow 0$.

4.2.4 A second example with infinite limits of integration

Let us convolve two Normal error functions which I shall express in the usual form showing the standard deviations in an explicit way; thus

$$\left.\begin{aligned} f(x) &= 1/(\sigma_1\sqrt{2\pi}) \exp(-x^2/2\sigma_1^2) \\ &\text{and} \\ g(y) &= 1/(\sigma_2\sqrt{2\pi}) \exp(-y^2/2\sigma_2^2) \end{aligned}\right\} \tag{4.23}$$

The convolution integral is

$$P(u) = 1/(2\pi\sigma_1\sigma_2) \int_{-\infty}^{\infty} \exp(-x^2/2\sigma_1^2) \exp\{-(u - x)^2/2\sigma_2^2\} dx$$

or

$$P(u) = 1/(2\pi\sigma_1\sigma_2) \int_{-\infty}^{\infty} \exp\{-(1/2\sigma_1^2 + 1/2\sigma_2^2)x^2 + ux/\sigma_2^2$$
$$- u^2/2\sigma_2^2\} \, dx$$

If we factor out of the exponential integrand that factor which is not a function of x we have

$$P(u) = 1/(2\pi\sigma_1\sigma_2) \exp(-u^2/2\sigma_2^2) \int_{-\infty}^{\infty} \exp\{-(1/2\sigma_1^2 + 1/2\sigma_2^2)x^2$$
$$+ ux/\sigma_2^2\} \, dx$$

If we equate the power of the new exponential integrand to the difference between a 'perfect square' and a suitable constant quantity we have

$$-(1/2\sigma_1^2 + 1/2\sigma_2^2)x^2 + (u/\sigma_2^2)x = -[(1/2\sigma_1^2 + 1/2\sigma_2^2x)$$
$$- (u/\sigma_2^2)/2\sqrt{(1/2\sigma_1^2 + 1/2\sigma_1^2)}]^2 + (u^2/\sigma_2^4)/4(1/2\sigma_1^2 + 1/2\sigma_2^2)$$

Now let

$$z = \sqrt{(1/2\sigma_1^2 + 1/2\sigma_2^2)}x - (u/\sigma_2^2)/2\sqrt{(1/2\sigma_1^2 + 1/2\sigma_2^2)}$$

whence

$$dx = dz/\sqrt{(1/2\sigma_1^2 + 1/2\sigma_2^2)}$$

and

$$P(u) = \frac{\exp(-u^2/2\sigma_2^2)}{2\pi\sigma_1\sigma_2\sqrt{(1/2\sigma_1^2 + 1/2\sigma_2^2)}} \int_{-\infty}^{\infty} \exp\{-z^2 + (u^2/\sigma_2^4)/4(1/2\sigma_1^2$$
$$+ 1/2\sigma_2^2)\} \, dz$$

or

$$P(u) = 1/\{\pi\sqrt{2}\sqrt{(\sigma_1^2 + \sigma_2^2)}\} \exp\{-u^2/2(\sigma_1^2 + \sigma_2^2)\} \int_{-\infty}^{\infty} \exp(-z^2) dz$$
$$= 1/(\sqrt{2\pi}\sqrt{(\sigma_1^2 + \sigma_2^2)}\} \exp\{-u^2/2(\sigma_1^2 + \sigma_2^2)\} \qquad (4.25)$$

Thus, the convolution integral of two Normal or Gaussian error functions has a variance that is the sum of the variances of the original functions, as stated in Note (c) below Table 2.1 in Chapter 2.

The reader should now have a thorough understanding of the limits of integration that apply to the convolution integral. When the range of variation of the two variables is between $-\infty$ and $+\infty$, the limits of integration become $-\infty$ and $+\infty$. When the range of one of the variables is a to b and the other is α to β where, say, $\alpha > a$ and $b > \beta$ the limits may be put at a and b, thus

$$\int_a^b$$

but great care must then be taken to interpret those limits properly when performing the actual integration.

5. The Delta Function

Figure 2.1 showed that convolving a given function $f(x_i)$ with a function consisting of a single ordinate of unit value at abscissa value y_i merely shifted the function $f(x_i)$ to the position $(x_i + y_i)$ thus translating it into the function $f(x_i + y_i)$. The example given in Section 4.2.3 stated that the Normal law of error

$$g(y) = (a/\sqrt{\pi}) \exp(-a^2y^2)$$

could be converted into a delta function or unit impulse of unit area and infinite value (at $0 - \epsilon \leqslant y \leqslant 0 + \epsilon$ where ϵ is a vanishingly small number). We had already ensured that the area under the Normal law was one, right at the beginning of the example and if we let $a \to \infty$ with $\epsilon \leqslant 1/a$ we have

$$g(\pm\epsilon) = (a/\sqrt{\pi})/e \leqslant 1$$

or, in the limit

$$g(0) \to \infty$$

In fact, all we need specify is that the product ϵa shall be finite or zero. As already stated, any function having the two properties (a) of unit area, (b) of infinite value at the zero value of abscissa and of zero value for all other abscissae, is called a delta function or a unit impulse and it is usually designated $\delta(x)$.

It is customary, because convenient, to drop one of the two variables x or y since, in mathematical convolution, the abscissae of the two functions have the same dimensions such as units of length, or time units, etc. We have already seen that our axes of abscissae in Figs. 4.2 to 4.7 have had to cope with x, y and u or x and u when y has been replaced by $u - x$. We can now confine our thinking to the single variable u and regard x as a variable of integration which enables us to integrate over certain defined ranges of the axis of abscissae measured in terms of u. In this way we may now write

$$\left.\begin{aligned} P(u) &= f(u)*g(u) \\ &= \int f(x)g(u-x)dx \\ &= \int f(u-x)g(x)dx \end{aligned}\right\} \qquad (5.1)$$

Note that I have omitted the limits of integration, because I haven't stated the valid ranges of variation of the abscissae of f and g.

Now consider

$$P(u) = f(u) * \delta(u) \tag{5.2}$$

We know from Fig.4.2 and the facts that $\delta(u)$ exists only when $u = 0$ and at that point its integral is one, that

$$P(u) = f(u + 0)$$
$$= f(u) \tag{5.3}$$

In other words

$$f(u) * \delta(u) = f(u) \tag{5.4}$$

It is quite easy to prove in an intuitive way. Thus

$$f(u) * \delta(u) = \int f(x)\delta(u - x)dx \tag{5.5}$$

Now $\delta(u - x)$ exists only when $x = u$, so we can write

$$f(u) * \delta(u) = \int f(u)\delta(0)dx$$

wherein we have substituted u for x, since the whole integral will be zero when $x \neq u$. Now $f(u)$ is not a function of x so we can extract it from the integral, writing

$$f(u) * \delta(u) = f(u) \int \delta(0)dx$$

$$= f(u)$$

because

$$\int \delta(0)dx = 1$$

Now try it the alternative way!

$$f(u) * \delta(u) = \int f(u - x)\delta(x)dx \tag{5.6}$$

$\delta(x)$ exists only when $x = 0$ so

$$f(u) * \delta(u) = \int f(u - 0)\delta(0)dx$$

$$= f(u)$$

Similarly[6]

$$f(u) * \delta(u - u_0) = f(u - u_0) \tag{5.7}$$

because

$$f(u) * \delta(u - u_0) = \int f(x)\delta(u - u_0 - x)dx$$

and

$\delta(u - u_0 - x)$ exists only when $x = u - u_0$

whence

$$f(u) * \delta(u - u_0) = f(u - u_0)\int \delta(0)dx$$

$$= f(u - u_0)$$

Equation 5.7 is very important in the study of transient mechanical and electrical phenomena, because it enables a function of time f(t) to be delayed by a time t_0 merely by convolving it with $\delta(t - t_0)$. Even more important, it allows us to express any waveform (of voltage or current or charge) or varying force function as a sum of delta functions of varying magnitudes occurring at successive instants of time throughout the duration of that waveform. This leads to the fundamental theorems of transient analysis that were stated by J.R. Carson[7] (1925), Campbell and Foster[8] (1927/28) and others. Their work was stimulated by the intuitive methods evolved by Oliver Heaviside[9] in the 1880s. The Laplace transforms, so popular today, constitute one of the integral-equation forms used, but in this author's opinion the Fourier integral is more straightforward and easier to understand; furthermore, it is directly applicable to optical-image functions and apertures which often have finite values on both sides of the mid-abscissa point and are therefore outside the range of validity of the Laplace integral.

6. Spectra and Characteristic Functions

In Chapter 2 we made use of generating functions as column markers. The generating function adopted was the simplest one having the required property of adding indices when multiplication was performed. Equation 2.7 in Chapter 2 defined that generating function. For reasons unknown to the author, generating functions used in connexion with continuous functions and mathematical convolution are called characteristic functions. The actual characteristic function adopted is not the smooth-function equivalent of

$$g(u) = \sum_{i=1} u^{x_i} f(x_i)$$

from Chapter 2, which in continuous-function form would be

$$g(u) = \int u^x f(x) dx \qquad (6.1)$$

but is, in fact, the Fourier transform of $f(x)$, thus

$$f[z] = \int f(x) \exp(-j2\pi zx) dx \qquad (6.2)$$

Many different notations are used to distinguish between a function, say $f(x)$ and its Fourier transform, but the one I have found most convenient for practical applications is to use square brackets, thus $f[z]$ is the Fourier transform of $f(x)$, and $f(x)$ is the inverse Fourier transform of $f[z]$. The Fourier series and the Fourier integral each convert a function of x into a different function whose independent variable has the dimension of $1/x$; thus, if $f(x)$ is a function of time measured in seconds, the $f[z]$ will be a function of frequency measured in hertz (cycles per second). If $f(x)$ is a function of length measured in millimetres, the $f[z]$ will be a function of a different kind of frequency measured in cycles per millimetre. Since the Fourier integral analyses the function $f(x)$ into an infinite number of sine waves and/or cosine waves, each of infinitesimal amplitude, it may be regarded as a spectrum analyser, and the function $f[z]$ is the spectrum whilst $f(x)$ is the synthesis of that spectrum. The reader will know that de Moivre's theorem shows that

$$\exp(-j2\pi zx) = \cos 2\pi zx - j \sin 2\pi zx$$

and therefore Equation 6.2 expresses the fact that $f[z]$ is the sum of an infinite number of cosine waves and sine waves in quadrature (because of the 'j'), each having the infinitesimal amplitude $f(x)dx$. The choice of the negative sign in the exponent of the exponential factor in the integral is arbitrary and conventional. $\exp(j\,2\pi zx)$ may be regarded, alternatively, as a rotating vector in the complex plane: $\cos 2\pi zx$ (abscissae), $j \sin 2\pi zx$ (ordinates). The magnitude or modulus will be

$$\sqrt{(\cos^2 2\pi zx + \sin^2 2\pi zx)} = 1$$

and the argument or angle through which the vector rotates when x varies from $x = 0$ to $x = x$ is $2\pi zx$ radians. Thus, $\exp(j\,2\pi zx)$ rotates in the positive geometric, or anticlockwise, direction and $\exp(-j\,2\pi zx)$ rotates in the negative geometric, or clockwise, direction. Which direction is chosen is arbitrary, but one cannot change direction of rotation once having made the choice as will be seen when we consider the inverse Fourier transform.

A spectrum may be considered in a general way, not merely as a set of electromagnetic waves such as the colours revealed from white light by a prism or the set of component waves that make up a given time-varying voltage or current. For instance, suppose $f(x)$ is a probability density function for which x is the error, measured in millimetres, in the lengths of mass-produced crankshafts. In this case $f[z]$ would be the spectrum with z measured in cycles of probability density per millimetre error and $f[z]$ (the ordinate values) in probability density per cycle per millimetre—hence 'waves of probability density' have a use if not a physical meaning!

Now, to return to characteristic functions. The reasons that the function $g(u)$ in Equation 6.1 is not used are that first, it is usually a difficult integral to solve and secondly and more important, are the combinations of very useful properties that the Fourier transform possesses. For example, if $f(x)$ is a probability density function, then its spectrum is the mean value of $\exp(-j2\pi zx)$, thus

$$f[z] = \overline{\exp(-j2\pi zx)} \qquad (6.3)$$

because, referring to Equation 2.1 in Chapter 2, and assuming the usual normalisation

$$\int f(x)dx = 1$$

the first-order moment of $\exp(-j2\pi zx)$ is given by Equation 6.2, just as in Equation 2.1

$$\bar{x} = \sum x_i f(x_i)$$

This leads us to see if the characteristic function $f[z]$ behaves in a similar way to the generating function $g(u)$ given by Equation 2.7. By differentiating $g(u)$ and putting $u = 1$, we found that (Equation 2.14a in Chapter 2)

$$g'(1) = M_1, g''(1) = M_2 - M_1$$

and continuing the process we have

$$M_1 = g'(1), M_2 = g''(1) + g'(1), M_3 = g'''(1) + 3g''(1) + g'(1)$$
$$M_4 = g''''(1) + 6g'''(1) + 7g''(1) + g'(1)$$

but no simple law seems to appear. Let us now consider successive derivatives of $f[z]$, thus

$$f'[z] = -j2\pi \int x f(x) \exp(-j2\pi zx) dx$$

$$f''[z] = (-j2\pi)^2 \int x^2 f(x) \exp(-j2\pi zx) dx$$

$$f^{(n)}[z] = (-j2\pi)^n \int x^n f(x) \exp(-j2\pi zx) dx$$

Now put $z = 0$ so that $\exp(-j2\pi zx) = 1$ and we have

$$f^{(n)}[z] = (-j2\pi)^n Mn \tag{6.4}$$

The *n*th derivative of the characteristic function is proportional to the *n*th moment of the original function, the coefficient of proportionality being $(-j2\pi)^n$

If we expand $f[z]$ in a MacLaurin series, thus

$$f[z] = f[0] + zf'[0] + \frac{z^2}{2!}f''[0] + \dots + \frac{z^n}{n!}f^{(n)}[0] + \dots$$

and replace the derivatives by the homologous moments, we have

$$f[z] = M_0 - j2\pi z M_1 + \frac{(j2\pi z)^2}{2!} M_2 - \frac{(j2\pi z)^3}{3!} M_3 + \dots \tag{6.5}$$

So, if the moments of a function $f(x)$ are known, the characteristic function can be calculated using Equation 6.5 and the function itself can then be found from the inverse Fourier transform[10]

$$f(x) = \int f[z] \exp(j2\pi xz) dz \tag{6.6}$$

If the range of validity of $f[z]$ is $-\infty < z < \infty$, then these values become the limits of integration in Equation 6.6. The characteristic function has a simpler relationship with the moments than does the generating function!

The reader who is unfamiliar with Fourier's theorems will wish to know the reason for the choice of the positive sign for the index of the exponential factor in the integrand in Equation 6.6, whereas the negative sign was chosen in Equation 6.2. Equation 6.6 gives $f(x)$ as a function of $f[z]$, but $f[z]$ was obtained as a function of $f(x)$ by Equation 6.2, thus the whole process is a sequence of operations, each subsequent operation depending upon what has gone before. We shall now show that, having adopted a negative exponent in calculating the spectrum $f[z]$, we must choose a positive exponent when calculating $f(x)$.

The Fourier integral theorem states that a function $f(x)$ can be analysed into a set of cosine and sine waves as follows:

$$\begin{aligned} f(x) &= \int_{-\infty}^{\infty} \int_{-\infty}^{\infty} f(u) \exp\{j2\pi z(x-u)\} \,du\,dz \\ &= \int_{-\infty}^{\infty} f(u) \exp(-j2\pi zu)\,du \int_{-\infty}^{\infty} \exp(j2\pi xz)\,dz \end{aligned} \tag{6.7}$$

where u is simply a variable of integration substituted in place of x to avoid confusion. The rules for double integration insist that the first, or left-hand, integral, when integrated, must then be inserted into the second integral, because it will be a function of z and the variable of integration for the second integral is z; however, Equation 6.7 can be broken down into two quite separate integral equations by giving a 'name' to the first integral. That 'name' is $f[z]$ and it turns out to be a very useful function, all on its own, for it describes a spectrum in mathematical notation. Thus, if we write

$$\int_{-\infty}^{\infty} f(u) \exp(-j2\pi zu)\,du = f[z]$$

and put this into the second integral, we have

$$f(x) = \int_{-\infty}^{\infty} f[z] \exp(j2\pi xz)\,dz$$

The proof of Equation 6.7 may be found in any number of books.[10-12] A physical interpretation is not too difficult, however. The existence of the Fourier series expansion.

$$\begin{aligned} f(x) = \tfrac{1}{2}b_0 &+ b_1 \cos x + b_2 \cos 2x + \ldots + b_n \cos nx + \ldots \\ &+ a_1 \sin x + a_2 \sin 2x + \ldots + a_n \sin {}_n x + \ldots \end{aligned}$$

is usually justified by showing that it is possible, if $f(x)$ is periodic, to calculate the coefficients b_n and a_n, and they turn out to have the following form

$$b_n = 1/\pi \int_{-\pi}^{\pi} f(u) \cos nu.\,du$$

$$a_n = 1/\pi \int_{-\pi}^{\pi} f(u) \sin nu.\,du$$

These are the weighting factors of the cosine and sine waves which, when added, give $f(x)$. When plotted as a series of ordinates against the successive values of n (the harmonic order), they form a line-spectrum. If the function $f(x)$ is not periodic, then by careful reasoning it can be imagined to have an infinite period and the Fourier series becomes the Fourier integral and the successive values of harmonic order n become infinitely close together, separated by $\Delta n \to dz$. In this situation and combining the sines and cosines into the exponential form by de Moivre's theorem, the combined coefficients b_n and a_n become f[z], which is a continuous spectrum instead of a line spectrum.

The reader who requires further mathematical insight into the Fourier integral will easily find many books to help him. The only reason for the superficial explanation given above is to aid the reader who may wish to apply it in practical problems such as the analysis of the response of linear electric circuits to transient excitations about which there are some further remarks to make, presently.

One further remark about the analogy between characteristic functions and generating functions. Let us consider the characteristic function f[z] from the viewpoint of position markers, as we did for the generating function g(u) given by Equation 2.7:

$$g(u) = \sum_{i=1}^{n} u^{x_i} f(x_i)$$

or its smooth-function equivalent from Equation 6.1

$$g(u) = \int u^x f(x) dx$$

The power of u, namely x, showed the column into which each term of a convolution should be placed. For the Fourier-transform type of characteristic function f[z], however, each function to be convolved is, as stated earlier, analysed into a set of sine and/or cosine waves; but once again, it is the power or index of exp $(-j2\pi zx)$ which is the column marker, only this time the 'columns' would be headed by the frequencies z of the suite of waves whose weighting factors or coefficients make up the characteristic function f[z]. When, in Equation 6.4, we put $z = 0$, we are then considering the zero-frequency values or d.c. components of the derivatives of the spectrum—not a very easy concept to grasp.

6.1 BOREL'S THEOREM

Before leaving this chapter there is a most important theorem to be studied. It plays a quite fundamental rôle in statistics and in the theory of the response of linear systems to transient excitations. Consider two functions f(x) and g(y). Let us calculate the convolution product (Equation 5.1)

$$P(u) = \int f(x) g(u - x) dx$$

where $u = x + y$.

The characteristic function or spectrum or Fourier transform of P(u) is, following Equation 6.2,

$$P[z] = \int \left\{ \int f(x) g(u - x) dx \right\} \exp(-j2\pi zu) du \tag{6.7}$$

or

$$P[z] = \iint f(x) g(u - x) \exp(-j2\pi zu) dx du$$

but

$$\exp(-j2\pi zu) = \exp(-j2\pi zx) \cdot \exp(-j2\pi zy)$$

and the variations of x are taken care of by the differential $\mathrm{d}x$, leaving the remaining variations of $\mathrm{d}u$ to be taken care of by $\mathrm{d}y$; thus we let $\mathrm{d}u = \mathrm{d}y$, whence

$$P[z] = \iint f(x) \exp(-j2\pi zx)\mathrm{d}x\, g(y) \exp(-j2\pi zy)\mathrm{d}y \qquad (6.8)$$

Since the first factor of the integrand is independent of y and the second factor is independent of x, the double integration may be separated into the product of two integrals

$$P[z] = \int f(x) \exp(-j2\pi zx)\mathrm{d}x \int g(y) \exp(-j2\pi zy)\mathrm{d}y \qquad (6.9)$$

whence

$$P[z] = f[z]g[z] \qquad (6.10)$$

The Fourier transform (or characteristic function) of a convolution product of two functions is the product of the Fourier transforms (or characteristic functions) of the functions that formed the convolution product.

This theorem can be generalised to cover convolution products containing more than two factors. Another interesting aspect is that analytic operations such as differentiating and integrating become algebraic operations such as multiplication and division of the characteristic functions. The simplifications that result can be very worth while indeed. In Chapter 8 we shall consider a few examples.

7. Fourier Transforms of the Delta Function

The next question to ask ourselves is 'what is the spectrum or characteristic function or Fourier transform of the delta function?' The answer is easily obtained; it is, from equation 6.2,

$$\mathrm{f}[z] = \int \delta(x) \exp(-j2\pi zx)\mathrm{d}x \quad (7.1)$$

Once again we remember that $\delta(x) = 0$ when $x \neq 0$; so putting $x = 0$ into Equation 7.1 we have

$$\left.\begin{aligned} \mathrm{f}[z] &= \int \delta(0)\mathrm{d}x \\ &= 1 \end{aligned}\right\} \quad (7.2)$$

So the spectrum is uniform, equal to one and independent of z, the frequency in cycles per unit of x, whatever that may be. So the single ordinate $\delta(x)$ at $x = 0$ in the $\mathrm{f}(x), x$ plane of co-ordinates becomes a uniform horizontal straight line in the $\mathrm{f}[z], z$ plane. Let us now consider the reverse question; namely, what is the inverse transform of the delta function in the $\mathrm{f}[z], z$ plane? Inserting $\delta[z]$ into Equation 6.6 gives

$$\mathrm{f}(x) = \int \delta[z] \exp(j2\pi xz)\mathrm{d}z \quad (7.3)$$

and remembering that $\delta(z) = 0$ when $z \neq 0$ and $\int \delta(0)\mathrm{d}z = 1$, we have $\mathrm{f}(x) = 1$, of course. What does this mean? It means that in letting $\mathrm{f}[z] = \delta[z]$ we have selected one of the complex $(\cos \pm j \sin)$ waves which, synthesised, make up $\mathrm{f}(x)$. The one we have selected is the one at zero frequency, $z = 0$, or in electrical terms, the d.c. component. Suppose we wanted to select a wave at frequency $z = \mathrm{a}$. We know that $\delta(z - a)$ will have unit area when $z = a$ and will be zero when $z \neq a$; thus the inverse transform of $\delta(z - a)$ is

$$\begin{aligned} \mathrm{f}(x) &= \int \delta(z - a) \exp(j2\pi xz)\mathrm{d}z \\ &= \exp(j2\pi xa) \int \delta(0)\mathrm{d}z \\ &= \exp(j2\pi xa) \end{aligned}$$

or

$$\mathrm{f}(x) = \cos 2\pi ax + j \sin 2\pi ax \quad (7.4)$$

which is a complex wave of frequency a, or a rotating vector of unit amplitude and angular velocity $2\pi a$ radians per unit of x.

Similarly, a delayed delta function $\delta(x - b)$ in the $f(x), x$ plane has the spectrum

$$f[z] = \int \delta(x - b) \exp(-j2\pi zx)dx \qquad (7.5)$$
$$= \exp(-j2\pi zb)$$

which is a complex spectrum of unit magnitude. The spectrum, being complex, requires a three dimensional representation, Fig. 7.1. Figure 7.1(b) shows $f[z]$ plotted as a helix with $2\pi bz$ as independent variable, whilst Fig. 7.1(a) shows the projection of the helix on a plane perpendicular to that of the paper. The points A, B, etc. are homologous on the two parts of the figure. The radius of the helix is, of course, the amplitude of $\exp(-j2\pi bz)$, namely one.

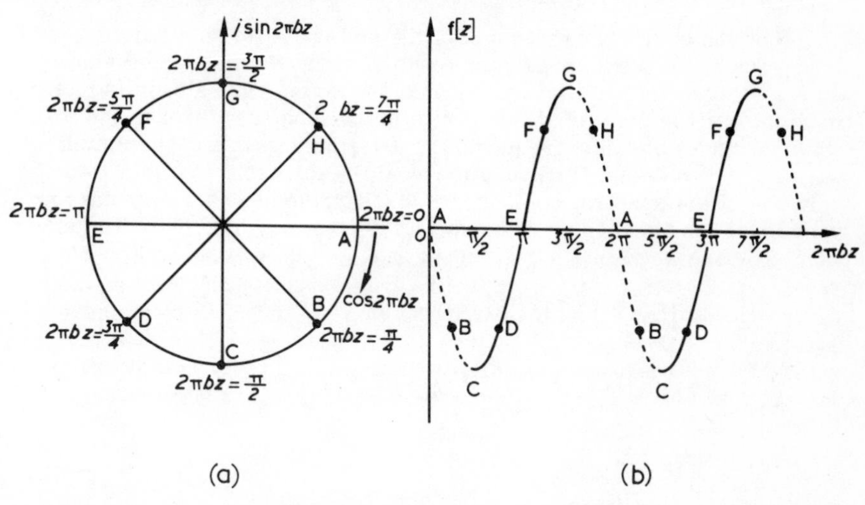

Fig. 7.1 Complex spectrum of a delayed delta function

7.1 AN ELECTRICAL EXAMPLE

The sampling property of the delta function is one of its principal uses. Let us consider a very simple example from electric-circuit theory. Consider an *RC* circuit as in Fig. 7.2. We have, as sinusoidal-voltage transfer-function

$$V_2/V_1 = \frac{-jX_C}{R - jX_C}$$

(where V_2 and V_1 are the complex values of sine wave output and input, respectively), considering the circuit as a potentiometer fed from a zero impedance generator of V_1 and connected to an open-circuit receiver at V_2. With $X_C = 1/\omega C$, we have

$$V_2/V_1 = 1/(1 + jRC\omega) \tag{7.6}$$

where $\omega = 2\pi f$ is the angular frequency in radians/second.

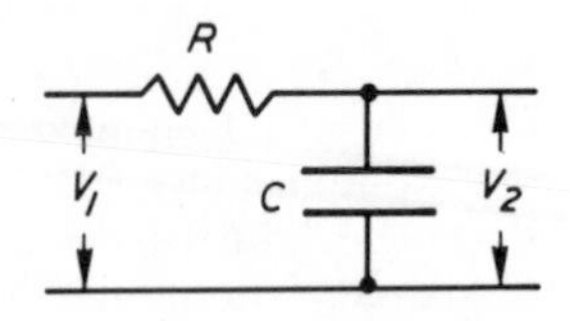

Fig. 7. 2 A resistance/capacitance circuit

Now the complex expression of the voltage ratio as a function of frequency is the weighting factor or coefficient of an assumed sinusoidal steady-state excitation that may be varied very slowly (so as to avoid all transient or shock-excitation phenomena) through the whole range of possible frequencies. It is, therefore, a spectrum. If we want to know what time function would result from the selection of only one of the spectral components (by, say, the use of a very narrow band ideal bandpass filter of bandwidth df) we may use the delta function. Suppose we select a component at $f_1 \pm \frac{1}{2}df$ we may write

$$\frac{v_2}{v_1}(t) = \int \delta[f - f_1] \frac{V_2}{V_1}[f] \exp(j2\pi t f) df$$

$$= \int \delta[f - f_1] \frac{\exp j2\pi t f}{1 + j2\pi RCf} df$$

$$= \frac{\exp(j2\pi t f_1)}{1 + j2\pi RCf_1} \tag{7.7}$$

We cannot, now, use the capital letter notation V_2/V_1 because we are no longer dealing with weighting factors or spectra, but with time functions, because we have now inserted the time-function excitation $\exp(j2\pi f t)$. If we multiply numerator and denominator by the conjugate $1 - j2\pi RCf_1$ of the denominator and expand the exponential excitation, we have

$$\frac{v_2}{v_1}(t) = \frac{1}{1 + (2\pi RCf_1)^2}[\cos 2\pi f_1 t + 2\pi RCf_1 \sin 2\pi f_1 t$$

$$+ j(\sin 2\pi f_1 t - 2\pi RCf_1 \cos 2\pi f_1 t)] \tag{7.8}$$

Now we must remember that, so far, the assumed excitation is $\cos 2\pi f_1 t + j \sin 2\pi f_1 t$. For a simple cosinusoidal excitation, therefore, we take the real part of Equation 7.8 and for a simple sinusoidal excitation we take the imaginary part; thus for $v_1 = V_1' \cos 2\pi f_1 t$, where V_1' is the amplitude,

$$\frac{v_2}{v_1}(t) = \frac{1}{\sqrt{\{1 + (2\pi RCf_1)^2\}}} \sin\,[2\pi f_1 t + \arctan\,(1/2\pi RCf_1)] \qquad (7.9)$$

If the capacitor is removed, $C = 0$ and $\arctan\,(1/2\pi RCf_1) = \pi/2$, whence

$$\frac{v_2}{v_1}(t) = \cos\, 2\pi f_1 t$$

which we should have expected.

If $v_1 = \sin\, 2\pi f_1 t$ we have

$$\frac{v_2}{v_1}(t) = \frac{1}{\sqrt{\{1 + (2\pi RCf_1)\}}} \sin\,[2\pi f_1 t + \arctan\,(-2\pi RCf_1)] \qquad (7.10)$$

Again, if $C = 0$, $\arctan\,(-2\pi RCf_1) = 0$ and

$$\frac{v_2}{v_1}(t) = \sin\, 2\pi f_1 t$$

which, again, we would have expected. This is an extraordinarily clumsy way of finding the response of an RC network to a sinusoidal (or cosinusoidal) excitation, but it does reveal the action of the spectral delta function as a 'frequency selecting' device. Of course, all electrical engineers know that if the transfer-function of a circuit is (Equation 7.6)

$$V_2/V_1 = 1/(1 + \mathrm{j}RC\omega)$$

$$= \frac{1 - \mathrm{j}RC\omega}{1 + (RC\omega)^2}$$

and the modulus or amplitude attenuation factor is

$$|V_2/V_1| = 1/\sqrt{\{1 + (RC\omega)^2\}}$$

and the phase shift undergone by a sinusoidal or cosinusoidal excitation is

$$\phi = \arctan\,(-RC\omega)$$

$$= -\arctan\,(RC\omega)$$

then, for an excitation $v_1 = V_1' \cos\,\omega_1 t$, the response is $v_2 = V_2' \cos\,(\omega_1 t + \phi)$ and for an excitation $v_1 = V_1 \sin\,\omega_1 t$ the response is $v_2 = V_2 \sin\,(\omega_1 t + \phi)$.

The reader will have observed that I have assumed a knowledge, on his part, of complex notation and steady-state (sinusoidal and cosinusoidal) circuit analysis. Incidentally, Campbell and Foster [8] use an elegant notation for the exponential wave $\exp\,(j\omega t)$; they call it cis ωt

for $\cos \omega t + i \sin \omega t$. That notation, put forward by Campbell in the Bell System Technical Journal in 1928, never 'caught on', but it remains available for those who like it.

I would now like to introduce the reader to the analysis of circuits to transient excitations. The word 'circuits' can be taken to mean 'linear dynamic systems' in general.

8. Response of Linear Circuits to Transient Excitations

8.1 FOURIER-INTEGRAL METHOD

We shall use only the Fourier integral rather than the Laplace integral or the Carson integral and since the best and most extensive tables of Fourier integrals are contained in *Fourier Integrals for Practical Applications*[8] we shall adopt the notation used in that book whilst retaining our own notation of parentheses for time or space functions and square brackets for frequency functions.

First, let us remind ourselves of the two Fourier transforms (Equations 6.2 and 6.6)

$$F[f] = \int G(g) \exp(-j2\pi fg)dg$$

and

$$G(g) = \int F[f] \exp(j2\pi gf)df$$

Here, f is frequency and g is time or distance or angle or even angular velocity, etc. In the above notation, either the new letter G or the use of both parentheses *and* brackets is redundant, but I shall continue to use both notations together, to help the reader. Campbell calls $F[f]$ and $G[g]$ 'paired coefficients', because if $F[f]$ be multiplied by $\exp(j2\pi tf)df$ and integrated, the function $G(t)$ will be obtained. Likewise $G(t)$ results from the convolution of $G(g)$ with $\delta(t)$, thus

$$G(t) = \int G(g)\delta(t-g)dg$$

So, if $G(g)$ be multiplied by $\delta(t-g)dg$ and integrated we again get $G(t)$. The function $F[f]$ may therefore be regarded as the coefficient of the cisoid $\exp(j2\pi tg)$ whilst $G(g)$ is the coefficient of the delta function $\delta(t-g)$. The tables in Reference 8 therefore, consist of two columns $F[f]$ and $G(g)$, each function being the Fourier transform of the other, with the positively rotating cisoid being attached to the coefficient $F[f]$ and the negatively rotating cisoid being attached to $G(g)$. In fact, I find this concept rather artificial and all the reader needs to remember is that g represents the independent time or space or, sometimes, angle variable and any letter of the alphabet may be substituted for it in accordance with the dictates of the problem to hand.

We shall now calculate the time response $R(t)$ of an electric circuit of transfer function $T[j2\pi f]$ to a transient excitation $E(t)$. An example of calculation of a transfer function has been given in Chapter 7.1 where we took the simple circuit of Fig. 7. 2. Had we calculated the ratio of the output voltage to the input current we should have called the transfer function a transfer impedance and the ratio of an output current to an input voltage would have been called a transfer admittance. All these functions, calculated in terms of resistance and reactance, indicate the response of the network to a sinusoidal (or cisoidal) excitation at any given frequency; they are therefore spectra and fulfil the requirement (a) of Section 4.1. Requirement (b), analyse the excitation function into a frequency spectrum of sinusoidal (or cisoidal) oscillations, can be achieved by the Fourier integral

$$E[f] = \int E(t) \exp(-j2\pi ft)dt \tag{8.1}$$

Requirement (c) is simply

$$R[f] = T[j2\pi f] \cdot E[f] \tag{8.2}$$

and requirement (d) is

$$R(t) = \int T(j2\pi f] \cdot E[f] \exp(j2\pi tf)df$$

or

$$R(t) = \int R[f] \exp(j2\pi tf)df \tag{8.3}$$

8.2 EXAMPLE: EXCITATION OF AN RC CIRCUIT BY AN EXPONENTIAL WAVEFORM

Let us use the simple *RC* circuit of Fig.7.2 and assume that the input voltage or excitation is

$$\left.\begin{aligned} E(t) &= \exp(-\alpha t) \text{ for } t \geqslant 0 \\ &\text{and} \\ E(t) &= 0 \text{ for } t < 0 \end{aligned}\right\} \tag{8.4}$$

The transfer function is given by Equation 7.6

$$\left.\begin{aligned} T[f] &= 1/(1 + jRC\omega) \\ &= 1/(1 + j2\pi RCf) \end{aligned}\right\}$$

Equation 8.1 applied to the function E(t) given by Equation 8.4 gives

$$\begin{aligned} E[f] &= \int_0^\infty \exp(-\alpha t) \cdot \exp(-j2\pi ft)dt \\ &= -1/(\alpha + j2\pi f) \mid \exp\{-\alpha + j2\pi f)t\} \Big|_0^\infty \\ &= 1/(\alpha + j2\pi f) \end{aligned} \tag{8.5}$$

So, from Equations 8.2, 7.6 and 8.5, we have

$$R[f] = \frac{1}{(1 + j2\pi RCf)\,(\alpha + j2\pi f)} \tag{8.6}$$

and, putting Equation 8.6 into Equation 8.3,

$$R(t) = \int_{-\infty}^{\infty} \frac{\exp\,(j2\pi tf)}{(1 + j2\pi RCf)\,(\alpha + j2\pi f)} \tag{8.7}$$

Once again, all limits of integration should be infinite unless, as was the case for the function $E(t)$, a definite restriction has been imposed; that was, $E(t)$, $= 0$ for $t < 0$. The integral 8.7 can be re-written as

$$R(t) = \int_{-\infty}^{\infty} \frac{e^{pt}}{(p + \alpha)\,(p + 1/RC)RC}\,df \tag{8.8}$$

where

$$p = j2\pi f$$
$$= j\omega$$

Integration of Equation 8.8 requires a knowledge of 'pole' and 'zero' manipulation, or it may be regarded as a contour integral requiring a knowledge of integration in the complex plane, and it was to avoid the necessity for such mathematics that Heaviside invented his 'operational calculus', but nowadays there are so many tables of integrals of this type that the engineer can use them with great benefit without having to acquire expert knowledge of advanced mathematics. I am not, of course, suggesting that such knowledge is not worth acquiring and I would advise the reader, now that he can see the utility of it, to acquire a working knowledge of poles and zeros and contour integration, but it is not essential for the solution of most practical problems, although it greatly aids understanding.

The 'coefficient' of $\exp\,(j2\pi gf)df$ in the integral 6.6, which will conform to Equation 8.8, is

$$F[f] = \frac{1/RC}{(p + \alpha)\,(p + 1/RC)} \quad (p = j2\pi f)$$

and the solution $G(g)$ is given in *Fourier Integrals for Practical Applications* as pair No. 448 in Table 1. Replacing g by t and $G(g)$ by $R(t)$ we have

$$R(t) = \frac{e^{-t/RC} - e^{-\alpha t}}{(\alpha - 1/RC)RC} \tag{8.9}$$

If the response of the circuit to a unit-step excitation

$$E(t) = 1 \text{ for } t \geqslant 0$$
$$E(t) = 0 \text{ for } t < 0$$

be required, all we have to do is to let $\alpha \to 0$, because $e^{-0} = 1$. Thus, for $E(t) = 1$, we have, from Equation 8.9,

$$R(t) = 1 - e^{-t/RC} \tag{8.10}$$

which is a well-known result. If the excitation were a delta function $E(t) = \delta(0)$, all we have to do is to note that the derivative with respect to time of a unit step is, in fact, a delta function and, since our *RC* circuit is a linear device, the response it produces to the derivative of a given excitation will be the derivative of the response to that excitation or, from Equation 8.10,

$$\frac{d}{dt} R(t) = (1/RC)e^{-t/RC} \tag{8.11}$$

Note that, if we assume the unit-step excitation, from which the unit impulse or delta function has been derived by differentiation, to be measured in volts, then the delta function will be in units of volts per second and its area will be measured in volts. If we excite dimensionless transfer function $T[f]$ with volts we shall get a response in volts, but if we excite it with volts/second we shall get a response in volts/second as shown by Equation 8.11, since the dimension of the product *RC* is time.

8.3 CONVOLUTION METHOD

This is described succinctly in Chapter 2 of Reference 4. We shall reproduce Woodward's method here, using our own notation.

Consider a network (electric circuit, for example) whose response to a delta function excitation $\delta(t)$ is $r(t)$. Its response to a 'bigger' or 'smaller' delta function $A\delta(t - \tau)$ at a later time τ will be $Ar(t - \tau)$, because we assume a linear circuit whose constant parameters are not dependent upon signal level. Now assume an excitation $E(t)$, which may be written, from equation 5. 5

$$\left.\begin{aligned} E(t) &= \int E(\tau)\delta(t - \tau)d\tau \\ E(t)*\delta(t) &= E(t) \end{aligned}\right\} \tag{8.12}$$

This means that we regard $E(t)$ as a succession of delta functions of magnitudes $E(\tau)d\tau$ occurring at times τ and separated by intervals $d\tau$. Now that refers to the input or excitation; but if the circuit responds to a delta function $A\delta(t - \tau)$ at instant τ by a response $Ar(t - \tau)$, we can replace $E(\tau)\delta(t - \tau)$ by $E(\tau)r(t - \tau)$ and finally, the response $R(t)$ becomes

$$R(t) = \int E(\tau)r(t - \tau)d\tau \tag{8.13}$$

or

$$R(t) = E(t)*r(t) \tag{8.14}$$

If we now compare Equation 8.13 with Equation 8.3 and we remember that Borel's theorem tells us that the Fourier transform of a product of two functions is equal to the convolution product of their Fourier transforms, we see that $E(t)$ is the transform of $E[f]$, which we already know, but also $r(t)$ must be the Fourier transform of $T[f]$. Thus the transfer function of a circuit, obtained by the usual complex impedance or admittance calculation well known to electrical engineers, is the Fourier transform of the time response of that circuit to a delta-function excitation. This is of capital importance as the reader will, no doubt, appreciate. Thus, with the aid of a table of Fourier transforms, we can calculate the time response of any circuit to a delta-function excitation if we know how to calculate its transfer function in terms of complex impedances or admittances. Use of the convolution integral of Equation 8.13 then gives the response of the circuit to any excitation $E(\tau)$.

8.4 EXAMPLE: EXCITATION OF AN RC CIRCUIT BY AN EXPONENTIAL WAVEFORM

We again use the circuit of Fig. 7.2 for which (Equation 7.6)

$$T[f] = 1/(1 + j2\pi RCf)$$

whence

$$T(t) = \int_{-\infty}^{\infty} T[f] \exp(j2\pi gf)\mathrm{d}f$$

$$= \int_{-\infty}^{\infty} \frac{\exp(j2\pi gf)\mathrm{d}f}{1 + j2\pi RCf} \tag{8.15}$$

$$= \int_{-\infty}^{\infty} \frac{\exp(\mathrm{gp})\,\mathrm{d}f}{1 + RCp} \quad \text{where } p = j2\pi f$$

The 'coefficient', $F[p]$, of $\exp(gp)\mathrm{d}f$ in the above integral is

$$F[p] = 1/(1 + RCp)$$

$$= \frac{1/RC}{1/RC + p}$$

and this is the 'frequency function' of pair No.438 in Table 1 of Reference 8. The time-function 'coefficient' of this pair is given as

$$G(g) = \frac{1}{RC}\,e^{-g/RC} \quad g \geqslant 0$$

so

$$T(t) = \frac{1}{RC}\,e^{-t/RC} \quad t \geqslant 0 \qquad [8.16)$$

which is in agreement with Equation 8.11. Having found the delta function response of our circuit, let us now calculate its response to the excitation (Equation 8. 4)

$$E(t) = e^{-\alpha t} \quad t \geq 0$$

Equation 8.14 gives

$$R(t) = E(t)*T(t) \quad \text{r being T, in this case.}$$

Equation 8.13 gives

$$R(t) = \int_0^t e^{-\alpha\tau} \frac{1}{RC} e^{-(t-\tau)/RC} d\tau \qquad (8.17$$

$$= \frac{e^{-t/RC}}{RC} \int_0^t e^{(1/RC - \alpha)\tau} d\tau$$

$$R(t) = \frac{e^{-t/RC} - e^{-\alpha t}}{(\alpha - 1/RC)RC}$$

which is the same result that was achieved by the Fourier integral method. (Equation 8. 9)

I hope the reader is now asking why the upper limit of integration is not ∞, as it was in Sections 4.2.3 and 4.2.4. The glib answer is that the cases treated in those two sections were convolution products whereas we are now dealing with a convolution function; that is, we are asking 'what is the convolution product of $E(t)$ with $T(t)$ up to a given time t?' Now the functions don't exist for negative time, but between $t = 0$ and $t = t$ they both have finite values and we require the integrated products of these values.

Consider a concrete example. Let

$$RC = 12.5 \text{ seconds}, \ 1/\alpha = 10 \text{ seconds}$$

whence

$$T(t) = (1/12.5)e^{-t/12.5} \text{ and } E(t) = e^{-t/10}$$

Equation 8.17 becomes

$$R(t) = \int_0^t e^{-\tau/10} \, (1/12.5)e^{-(t-\tau)/12.5} d\tau$$

Now let us suppose we want the response at a given fixed time $t_1 = 10$ seconds

$$R(t_1) = 1/12.5 \int_0^{t_1} e^{-\tau/10} e^{-(t_1-\tau)/12.5} d\tau \qquad (8.18)$$

Figure 8.1 shows $e^{-\tau/10}$, $e^{-\tau/12.5}$ and, (between $\tau = 0$ and $\tau = t_1$), $e^{-(\tau_1-\tau)/12.5}$. The integral above thus refers to the interval

$0 < \tau < t_1$. Once again, we see that it is the limits of integration which require the most careful thought. Equation 8.18 becomes, of course, the same as Equation 8.9 with $t = t_1 = 10$; thus

$$R(10) = 1/12.5 \frac{e^{-10/12.5}e^{-10/10}}{1/10 - 1/12.5}$$

$$= 0.328$$

and this is the value of R(t) for the specific time epoch $t = 10$ seconds. The variable of integration τ, by exploring the time interval from switch-on at $t = 0$ up to the time t, at which we wish to know the response, is investigating the performance of the circuit for all relevant past times. This is necessary, because the response of any circuit, except one containing only resistance, depends on its past 'history'. In fact, the function T(t) has been called the 'coefficient of heredity' by some mathematicians.

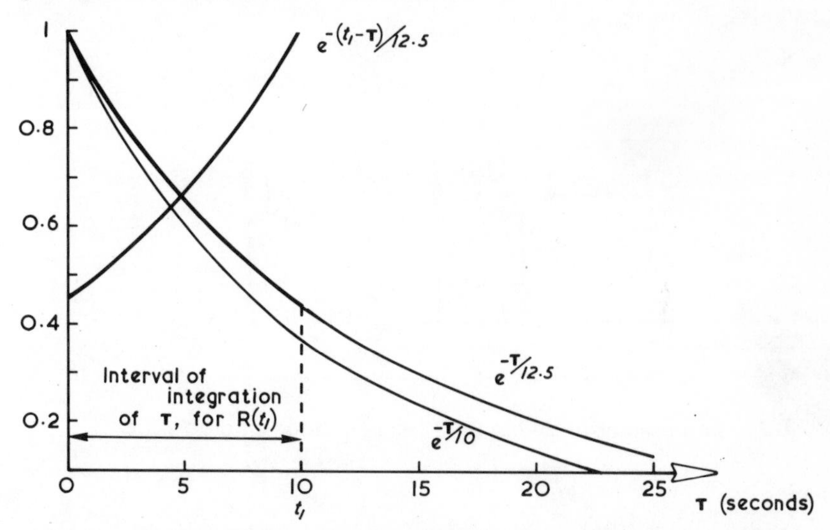

Fig. 8.1 Response of an *RC* circuit to an exponential excitation

A chapter dealing with the response of linear circuits to transient excitations would be incomplete without a brief glance at one of the more erudite—but not the less important or practical for that reason—problems for which the solutions requiring advanced mathematical treatment are already known. I refer to the response of electrical transmission lines to transient excitations. The general case of a transmission line of length l terminated with an impedance Z_T with one or more impedances across it at points x_1, x_2, etc. along it and fed through an impedance Z_0 is far too complicated to be dealt with here. We shall consider a relatively simple case which, however, will reveal one or two interesting properties and will require an exemplary use of the tables of 'paired coefficients' in Reference 8.

8.5 EXAMPLE[10]: A TRANSMISSION LINE TERMINATED BY AN IMPEDANCE Z_T

Consider a smooth transmission line in which each single section of infinitesimal length dx contains a series impedance dz and a shunt admittance dA, Fig. 8.2. The length of the line is l and it is terminated at the far, or output, end by an impedance $Z_T[p]$ where $p = j\omega$ or $j2\pi f$ as before. The voltage drop across dz is

$$-\mathrm{d}v = i\mathrm{d}z$$

and, neglecting the diminution dv of the voltage v across dA, we find for the current drain through dA

$$-\mathrm{d}i = v\mathrm{d}A$$

Now, if L and R are the series inductance and resistance per unit length dx and G and C are the shunt conductance and capacitance per unit length dx we have

$$\mathrm{d}z = (R + pL)\mathrm{d}x$$

$$\mathrm{d}A = (G + pC)\mathrm{d}x$$

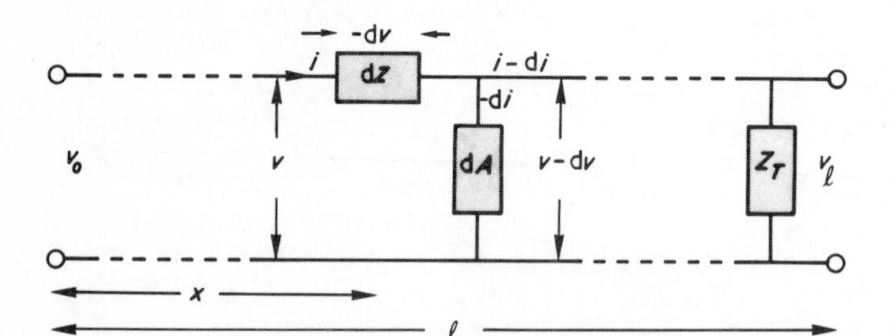

Fig. 8.2 An element of an electric transmission line

whence

$$\mathrm{d}v/\mathrm{d}x = -i(R + pL) \tag{8.19}$$

$$\mathrm{d}i/\mathrm{d}x = -v(G + pC) \tag{8.20}$$

and differentiating each equation with respect to x we have

$$\mathrm{d}^2v/\mathrm{d}x^2 = -\mathrm{d}i/\mathrm{d}x(R + pL) \tag{8.21}$$

$$\mathrm{d}^2i/\mathrm{d}x^2 = -\mathrm{d}v/\mathrm{d}x(G + pC) \tag{8.22}$$

If we put di/dx from equation 8.20 into equation 8.21 and dv/dx from equation 8.19 into equation 8.22 we arrive at

$$\mathrm{d}^2v/\mathrm{d}x^2 = v(G + pC)(R + pL)$$

and

$$\mathrm{d}^2i/\mathrm{d}x^2 = i(R + pL)(G + pC)$$

or, if we let

$$\gamma^2 = (G + pC)(R + pL)$$

we have

$$d^2v/dx^2 - \gamma^2 v = 0 \tag{8.23}$$

$$d^2i/dx^2 - \gamma^2 i = 0 \tag{8.24}$$

These are very well known linear differential equations with constant coefficients. They tell us that, taking Equation 8.23 for example, v is a function of x such that its second derivative is proportional to itself, the constant of proportionality being γ^2. It can easily be shown that, if a and b are constants to be determined by the boundary conditions

$$v = ae^{\gamma x} + be^{-\gamma x} \tag{8.25}$$

is a solution, but to show that it is the only solution is outside the scope of this book. Incidentally, linear differential equations can be solved by Fourier integral methods, but in our case the boundary conditions are rather complicated and it is simpler to deal with them by classical means.

Now, since γ is a function of $p = j2\pi f$, we note that v is a function of both p and x. First, we settle the boundary conditions by considering what happens at the input and output of the transmission line whilst considering $v(p, x)$ as a spectrum in terms of f or p and examining its variation as a function of x, the distance along the line. Later we shall attend to the effects upon v of variation of p. Rather than introduce two further constants like a and b, we shall calculate the current, not by solving Equation 8.24, but by calculating dv/dx from Equation 8.25, thus

$$dv/dx = \gamma(ae^{\gamma x} - be^{-\gamma x})$$

and inserting this value of dv/dx into Equation 8.19. We thus get

$$i = -\gamma/(R + pL) \cdot (ae^{\gamma x} - be^{-\gamma x})$$

$$= \sqrt{\{(Cp + G)/(R + pL)\}} \cdot (be^{-\gamma x} - ae^{\gamma x}) \tag{8.26}$$

Now for the boundary conditions which will determine the constants a and b!

Let us excite the line with an input voltage having the spectrum $v_0[p]$. So, for $x = 0$, Equation 8.25 becomes

$$v_0[p] = a + b \tag{8.27}$$

At the end of the line, where $x = l$, we have

$$V_l[p] = i_l[p]Z_T[p] \tag{8.28}$$

so Equation 8.26 for the current becomes

$$i_l[p] = \sqrt{\{(G + Cp)/(R + pL)\}} \cdot (be^{-\gamma l} - ae^{\gamma l}) \tag{8.29}$$

Putting i from Equation 8.29 into Equation 8.28 we have for $x = l$

$$v_l[p] = Z_T[p]\sqrt{\{(G + pC)/(R + pL)\}} \cdot (b\mathrm{e}^{-\gamma l} - a\mathrm{e}^{\gamma l}) \qquad (8.30)$$

But from Equation 8.25 with $x = l$ and Equation 8.30 we have

$$a\mathrm{e}^{\gamma l} + b\mathrm{e}^{-\gamma l} = Z_T[p]\sqrt{\{(G + pC)/(R + pL)\}} \cdot (b\mathrm{e}^{-\gamma l} - a\mathrm{e}^{\gamma l}) \qquad (8.31)$$

Now let

$$Z_T[p]\sqrt{\{(G + pC)/(R + pL)\}} = \lambda$$

whence, from Equation 8.27 and

$$a\mathrm{e}^{\gamma l} + b\mathrm{e}^{-\gamma l} = \lambda(b\mathrm{e}^{-\gamma l} - a\mathrm{e}^{\gamma l}) \qquad (8.32)$$

we derive

$$\left.\begin{aligned} a &= v_0[p]\,\frac{(\lambda - 1)\mathrm{e}^{-\gamma l}}{(\lambda + 1)\mathrm{e}^{\gamma l} + (\lambda - 1)\mathrm{e}^{-\gamma l}} \\ b &= v_0[p]\,\frac{(\lambda + 1)\mathrm{e}^{\gamma l}}{(\lambda + 1)\mathrm{e}^{\gamma l} + (\lambda - 1)\mathrm{e}^{-\lambda l}} \end{aligned}\right\} \qquad (8.33)$$

Remembering that

$$\mathrm{e}^{\gamma l} + \mathrm{e}^{-\gamma l} = 2\cosh\gamma l$$

$$\mathrm{e}^{\gamma l} - \mathrm{e}^{-\gamma l} = 2\sinh\gamma l$$

and putting the values of a and b from Equations 8.33 into Equation 8.25 we have, finally, for all values of x

$$v[p, x] = \frac{(\lambda + 1)\mathrm{e}^{\gamma(l-x)} + (\lambda - 1)\mathrm{e}^{-\gamma(l-x)}}{2(\lambda\cosh\gamma l + \sinh\gamma l)} \qquad (8.34)$$

The mathematical manipulation required to reach the above result is complicated, but not difficult. The mathematics required to transform Equation 8.34 into a transient time function $v(t)$, given a transient excitation $v_0(t)$, is indeed difficult, but several specific examples are known and may be found in Reference 8 and elsewhere. First, however, let us simplify Equation 8 34 by changing the variable x from being a measure of the distance from the input to a measure of the distance back, away from, the output. Thus $l - x$ becomes y and

$$v[p, y] = \frac{\lambda\cosh\gamma y + \sinh\gamma y}{\lambda\cosh\gamma l + \sinh\gamma l}\,V_0[p] \qquad (8.35)$$

with

$$\gamma = \sqrt{\{(G + pC)\,(R + pL)\}}$$

and

$$\lambda = Z_T\sqrt{\{(G + pC)/(R + pL)\}}$$

Similarly, the current flowing along the line, at a point y from the output or Z_T end, is

$$i[p, y] = (\lambda / Z_T) \frac{\lambda \sinh \gamma y + \cosh \gamma y}{\lambda \cosh \gamma l + \sinh \gamma l} \, v_0[p] \tag{8.36}$$

Note that γ has the dimension 1/length, because G, C, R and L are measured in the appropriate electrical units *per unit of length*. On the other hand λ is dimensionless. In order to reveal some of the physical phenomena that take place along the line we make two considerable simplifications. First, we let $\lambda = 1$, which means that we let

$$Z_T = \sqrt{\{(R + pL)/(G + pC)} \tag{8.37}$$

and secondly, we let $R/L = G/C$ whence

$$\gamma = (p + R/L)\sqrt{LC} \tag{8.38}$$

The effect of Equation 8.37 is to transform Equations 8.35 and 8.36 into

$$v[p, y] = v_0[p]\mathrm{e}^{-\gamma(l-y)} \tag{8.35a}$$

$$i\,[p, y] = (v_0[p]/Z_T)\mathrm{e}^{-\gamma(l-y)} \tag{8.36a}$$

$l - y$ is, of course, the distance away from the input end of the line, namely x, so we re-write Equation 8.35a and 8.36a as

$$v[p, x] = v_0[p]\mathrm{e}^{-\gamma x} \tag{8.35b}$$

$$i[p, x] = (v_0[p]/Z_T)\mathrm{e}^{-\gamma x} \tag{8.36b}$$

These show that the voltage and current diminish as x increases, but as γ is a complex quantity we shall have to examine the matter in greater detail. Remembering that $p = j2\pi f = j\omega$, we may write

$$v[j\omega, x] = v_0[j\omega]\mathrm{e}^{-Rx\sqrt{(C/L)}} \cdot \mathrm{e}^{-j\omega x\sqrt{(LC)}} \tag{8.35c}$$

$$i[j\,\omega, x] = (v_0[j\,\omega]/Z_T)\mathrm{e}^{-Rx\sqrt{(C/L)}} \cdot \mathrm{e}^{-j\omega x\sqrt{(L/C)}} \tag{8.36c}$$

8.5.1 Steady-state solution

The ratios v/v_0 and i/i_0 may be regarded in the same light as the ratio V_2/V_1 of Equation 7.6; that is, they are complex transfer functions; so when v_0 takes the form of a cisoidal excitation

$$v_0 = V_0\mathrm{e}^{j\,\omega t} \tag{8.39}$$

v may be calculated for any distance x along the line. Thus, we have

$$v(t, x) = V_0'\mathrm{e}^{j\omega t}\mathrm{e}^{-Rx\sqrt{(C/L)}}\mathrm{e}^{-j\omega x\sqrt{(LC)}}$$

$$= V_0'\mathrm{e}^{-Rx\sqrt{(C/L)}}\mathrm{e}^{j\omega(t-x\sqrt{(LC)}} \tag{8.40}$$

and if the excitation is, for example, a cosine wave, we must take the real part of Equation 8.40, thus

$$v(t, x) = \mathrm{V}_0'\mathrm{e}^{-Rx\sqrt{(C/L)}} \cos \omega\{t - x\sqrt{(LC)}\} \tag{8.41}$$

So the amplitude $V_0'\mathrm{e}^{-Rx\sqrt{(C/L)}}$ of the wave diminishes exponentially as x increases and the phase $\omega(t - x\surd LC)$ of the wave increases with t and decreases with x. The speed of the phase of the wave may be found thus:

$$\phi = \omega t - \omega x \surd(LC)$$

whence

$$x = (\omega t - \phi)/\omega\surd(LC)$$

and the phase speed is

$$dx/dt = 1/\surd(LC)$$

which never quite reaches the speed of light or electromagnetic waves in vacuo. Although this happens to be the speed of propagation of waves along the line in the special case we have cited (as we shall see later), namely when $\lambda = 1$ or $R/L = G/C$, dx/dt is, in this case, merely the speed of the phase of the wave; that is, if we examined the phase ϕ_1 at a given time t_1 and distance x_1 and then we moved our phase-measuring probe along the line at speed $1/\surd(LC)$ we should continually measure that particular value ϕ_1 of phase.

Equations 8.40 and 8.41 are steady-state solutions in that the cisoidal excitation is assumed to be continuous; that is, switched on at $t = -\infty$ so that all transient responses that were due to the switching on process have died away long ago. It is, in my view, nonsense to talk about a speed of propagation of a periodic function which has existed from $t = -\infty$ and will continue until $t = +\infty$, for what attribute of the function can we take in order to follow its propagation along the line? Incidentally, Equation 8.40 shows that there is only one wave present in the line and it becomes increasingly attenuated towards the output end where Z_T is connected. This is because we made $\lambda = 1$ and thereby got rid of the second term in the numerator of Equation 8.34. That second term describes a wave whose attenuation increases as x decreases; that is, it is the reflection of the original wave at the terminating impedance Z_T. By giving Z_T the value shown by Equation 8.37, the reflected wave is eliminated, because the incident wave can dissipate itself in Z_T thus behaving as if the line continued forever. This special value of Z_T is called the characteristic impedance of the line and is, of course, of very great importance.

The other simplifying assumption that we made was that shown by Equation 8.38. This enabled us to remove p from the square root implicit in the definition of γ. The physical effect of this is to render the line distortionless, because although the amplitude is progressively attenuated along the line, it is independent of frequency, $\omega/2\pi$. In the more general case where

$$\gamma^2 = (G + j\omega C)(R + j\omega L)$$

the expression for the steady-state voltage at time t and distance x along the line becomes

$$v(t, x) = V_0' e^{-\gamma x} \cdot e^{j\omega t}$$

$$= V_0 \exp\left[-\sqrt{\{(G + j\omega C)(R + j\omega L)x\}}\right] \exp j\omega t \quad (8.42)$$

In converting the square-rooted factor of x in the first exponential into a straightforward complex quantity $a(\omega) + jb(\omega)$ we find that both a and b contain square and fourth roots of terms containing powers of ω up to the fourth; thus, the response of the line is no longer independent of frequency. For the interested reader, the exponent of the first exponential in Equation 8.42 becomes

$$\left.\begin{array}{l} -1/\sqrt{2}\,\{[(\omega^2 LC + RG)^2 - \omega^2(RC - LG)^2]^{1/2} + RG - \omega^2 LC\}^{1/2} x \\ -j/\sqrt{2}\,\{[(\omega^2 LC + RG)^2 - \omega^2(RC - LG)^2]^{1/2} - RG + \omega^2 LC\}^{1/2} x \end{array}\right\} \quad (8.43)$$

So the attenuation of the distorted wave is indicated by the first term of Expression 8.43 whilst the term giving rise to the wave in time and space is the second one in Expression 8.43. For a given frequency of input excitation $\omega_1/2\pi$ it is complicated, but not difficult to calculate the voltage at any point x along the line.

8.5.2 Transient solution for delta-function excitation

Now, having given a brief discussion about the steady-state performance, let us examine the transient performance. We shall then be able to talk about a real speed of propagation—if there is one!

Let us consider the simple case, first. Thus we revert to Equation 8.35c, but now, instead of taking a cisoidal excitation $v_0 = V_0' {}^{j\omega t}$ we will choose a suitable transient. Why not choose our old friend, the delta function? We know that the spectrum of the delta function is constant and equal to unity, so let's choose a delta function of area or value V_0'. Thus Equation 8.35c now becomes

$$v[j\omega, x] = V_0' e^{-Rx\sqrt{(C/L)}} \cdot e^{-j\omega x\sqrt{(LC)}} \quad (8.44)$$

But, $\exp\{-j2\pi f x\sqrt{(LC)}\}$ is the spectrum of a delta function occurring at the delayed time $t - x\sqrt{(LC)}$, as can be seen from Equation 7.5 if we put, from that equation, $z = f$, $b = x\sqrt{(LC)}$ and $x = t$. Thus

$$v(t, x) = V_0' e^{Rx\sqrt{(C/L)}} \delta\{t - x\sqrt{(LC)}\} \quad (8.45)$$

Remembering that the delta function exists only when $t = x\sqrt{(LC)}$ we can see that the voltage along the line becomes an attenuated delta function moving at the speed of

$$\begin{aligned} dx/dt &= (d/dt)\{t/\sqrt{(LC)}\} \\ &= 1/\sqrt{(LC)} \end{aligned}$$

because the voltage impulse $\delta\{t - x\sqrt{(LC)}\}$ exists only when $t = x\sqrt{(LC)}$. Incidentally, Equation 8.45 is valid for all values of t, but it is zero except when $t = x\sqrt{(LC)}$ when it has the value

$$v(t,x) = V_0' e^{-Rx\sqrt{(C/L)}}$$
$$= V_0' e^{-Rt/L}$$

The whole phenomenon is rather analogous to a speeded-up version of a python swallowing a pig; the pig becoming progressively more attenuated as it slides further along the python's body!

8.5.3 Transient solution for unit-step excitation

Suppose, now, that the excitation is a unit step of voltage,

$$v_0(t) = V_0' \text{ for } t \geqslant 0$$

and

$$v_0(t) = 0 \text{ for } t < 0$$

Just as we noted before Equation 8.11 in Section 8.2 that the delta function is the derivative of the unit step, so we can say that the unit step is the integral of the delta function. Thus

$$v(t,x) = V_0' e^{-Rx\sqrt{(C/L)}} \int_0^{0>t>\infty} \delta(t-0)\mathrm{d}t$$

$$v(t,x) = V_0' e^{-Rx\sqrt{(C/L)}} \tag{8.46}$$

$\delta(t-0)$, because $x\sqrt{(LC)} = 0$ at the point of excitation.

8.5.4 Transient solution for exponential excitation

Let $v_0(t) = V_0' e^{-\alpha t}$

Since we already have the response of the line to a delta-function excitation (Equation 8.45) we can use the convolution method as shown by Equation 8.13 in Section 8.3; whence

$$v(t,x) = V_0' e^{-Rx\sqrt{(C/L)}} \int_0^{0<t<\infty} e^{-\alpha\tau}\delta\{t - x\sqrt{(LC)} - \tau\}\mathrm{d}\tau$$

Since $\delta(t - x\sqrt{(LC)} - \tau)$ differs from zero only when $\tau = t - x\sqrt{(LC)}$ we have

$$v(t,x) = V_0 e^{-Rx\sqrt{(C/L)}} e^{-\alpha\{t - x\sqrt{(LC)}3} \tag{8.47}$$

Equations 8.45, 8.46 and 8.47 giving the response of the transmission line to delta-function, unit-step and exponential excitations have each been found by our own direct labours, but we could have found them immediately from Reference 8. Equation 8.44 with p written instead of $j\omega$ appears in the second column of Table II as item 10 and its time-function transform, equation 8.45, is given in the third column, with the delta function $\delta\{t - x\sqrt{(LC)}\}$ written as $S_0(t - x/v)$ where v is the velocity of propagation. Similarly, the unit-step case, equation 8.46, appears in the fourth column of item 10 in Table II. The case of the exponential excitation e^{-at} can be obtained by finding the spectrum or transform of it as pair no.438 in Table I where we find that it transforms to $1/(p + \alpha)$. If we now multiply the spectrum in Equation 8.44 by $1/(p + \alpha)$ we obtain the spectrum of pair no. 604 in Table 1. The transform of this spectrum is given as in Equation 8.47.

The more complicated cases for which our simplifying assumptions, (Equations 8.37 and 8.38), do not hold may also be found by judicious use of the above-mentioned tables.

The reader will, by now, have appreciated that sometimes it is easier to use the Fourier integral method throughout a problem, but sometimes the convolution integral suggests itself. It may largely depend on what kind of tables of integrals the engineer will have before him, readily available. The power of the delta function and the relative simplicity of its uses should be appreciated. It is the finest 'test signal' yet devised and if its properties are well understood it often leads to relatively simple methods using convolution rather than Fourier integrals.

9. Some of the Rules for Playing the Fourier-Transform Game

We shall use the notation $p = j2\pi f$ frequently for reasons that will become evident. We shall also use the letter g instead of t (for time) in order to conform to the notation used in Reference 8.

Rule (i): Table 1, pair 208

$$p \doteqdot d/dg \text{ where } \doteqdot \text{ means 'transform of'} \tag{9.1}$$

Re-write Equation 6.6 as

$$G(g) = \int F[f]e^{pg}df$$

Now differentiate with respect to g, thus

$$\frac{dG}{dg} = \int F[f]\frac{d}{dg}\,e^{pg}df$$

$$= \int F[f]pe^{pg}df$$

So, differentiating G with respect to g is equivalent to the operation of multiplying F by p. *Heaviside's Operational Calculus* is based upon this. It is easy to see that

$$1/p \doteqdot \int dg \tag{9.2}$$

and, furthermore, that

$$p^n \doteqdot d^n/dg^n \text{ and } 1/p^n \doteqdot \overset{1\,2}{\iint} \cdots \overset{n}{\int} dg^n \tag{9.3}$$

These facts follow from the adoption, for spectrum analysis, of a periodic function that, apart from a multiplying factor, reproduces itself on differentiation and on integration. Thus, for example,

$$(d/dt)e^{-\alpha t} = -\alpha e^{-\alpha t} \quad \int e^{-\alpha t}dt = -1/\alpha e^{-\alpha t}$$

$$(d^2/dt^2)e^{-\alpha t} = \alpha^2 e^{-\alpha t} \quad \iint e^{-\alpha t}dt^2 = 1/\alpha^2 e^{-\alpha t}$$

$$(d^3/dt^3)e^{-\alpha t} = -\alpha^3 e^{-\alpha t} \quad \iiint e^{-\alpha t}dt^3 = -1/\alpha^3 e^{-\alpha t}$$

and so on.

Since all linear differential equations with constant coefficients are solved by letting the unknown function or dependent variable take the form of a sum of exponentials, we see that Rule (i) has a large field of application. On the spectrum side of the sign ÷ we can conceive of quantities like $\sqrt{p}$ which involve advanced concepts of differentiation and integration, which we cannot go into here.

Rule (ii): Table 1, pair 205

$$G(g/a) \doteqdot a \cdot F[af] \qquad (9.4)$$

Again, take Equation 6.6, but let the independent variable of the spectrum $F[f]$ be af instead of simply f. Thus,

$$\begin{aligned} G(g) &= \int F[af] e^{j2\pi gf} df \\ &= \frac{1}{a} \int F[af] e^{j2\pi(g/a)af} daf \\ &= \frac{1}{a} \int F[f] e^{j2\pi fg/a} df \end{aligned} \qquad (9.5)$$

if we substitute f for af. But, Equation 9.5 is the same as $(1/a)G(g/a)$.

Rule (iii): Table I, pair 201

$$F_1[f] + F_2[f] \doteqdot G_1(g) + G_2(g) \qquad (9.6)$$

Rule (iv): Table I, pair 207

$$G(g - g_0) \doteqdot e^{-pg_0} F[f] \qquad (9.7)$$

Electrical engineers will know this one. Consider a cisoidal voltage $V_0' e^{j\omega t}$. Now multiply it by $e^{-p\tau} = e^{-j\omega\tau}$ we get

$$V_0' e^{j\omega t} \cdot e^{-j\omega\tau} = V_0' e^{j\omega(t - \tau)}$$

That is, we have retarded the phase ωt by an amount $\omega\tau$ or delayed the waveform by a time τ. Of course, e^{-pg_0} is the spectrum of a delayed delta function, as we have already seen, and the convolution of $G(g)$ with $\delta(g - g_0)$ yields $G(g - g_0)$.

Rule (v): Table I, pair 202

$$F_1[f] \cdot F_2[f] \doteqdot \int G_1(x) \cdot G_2(g - x) dx \qquad (9.8)$$

This is Borel's theorem as expounded in Section 6.1.

There are many more rules, but it is hoped that mention of the above five of them will whet the reader's appetite for more.

10. Examples of Use of the Fourier Integral and Convolution in Broadcasting Problems

10.1 TRANSIENT BEHAVIOUR OF A TELEVISION RECEIVER 'FLYWHEEL' CIRCUIT[13]

The scanning of the electron beam in the display tube of a television receiver is synchronised to the information contained in the transmitted picture signal by means of rectangular pulses, one at the commencement of each scanning line and a block of similar pulses at the commencement of each field; two fields whose lines are interlaced constitute one complete picture. The line-synchronising pulses are repeated at a frequency of 15 625 Hz, that is, of course, 625 lines in a twenty-fifth of a second. Unwanted, but inevitable random 'noise' can disturb the synchronisation as those who live beyond the fringe of a television service area will know. In order to minimise disturbance due to random noise, the circuits that use the line-synchronising pulses for controlling the scanning are made to have a sluggish response to any changes or disturbances to the line-synchronising-pulse frequency; that is, they are given a narrow bandwidth around the line-synchronising frequency of 15 625 Hz. This renders the receiver scanning circuits less responsive to disturbances than would otherwise be the case. Now, however, suppose that due, for example, to the use of certain types of video-tape recorders, the stability of repetition of the television signal including its line-synchronising pulses is less than perfect. The receiver scanning circuits will be unable to follow the 'jittering' signals to an extent which will depend upon both the magnitude of the jitter and its frequency. Assuming a given receiver line-scan circuit, what values of tolerance should be applied to the magnitude and frequency of transmitted disturbance so as not to annoy the viewer? The first thing to do is to conduct tests (in the laboratory) with 'guinea-pig' viewers to see how much disturbance to a television image can be tolerated.
The second thing is then to calculate the limiting amount of transmitted-signal jitter that, after attenuation through the receiver flywheel circuit, just equals the tolerable amount.

A typical flywheel circuit[14] is shown schematically in Fig. 10.1. Incoming line-sync pulses at the nominal frequency f_0, perturbed by

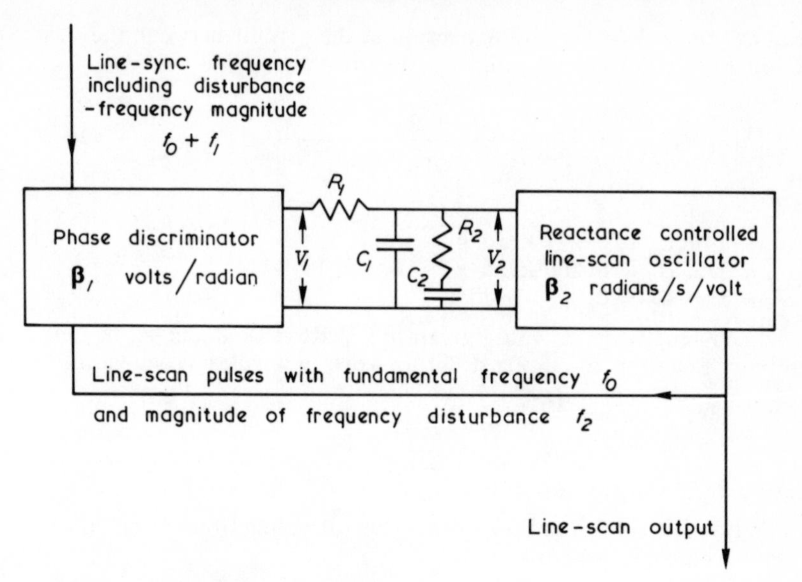

Fig. 10.1 'Flywheel' circuit for line-scanning in a television receiver

an amount (in frequency) f_1 are fed to a phase discriminator. At the same time, a portion of the output voltage, nominally also at frequency f_0, but now perturbed in frequency by f_2 is fed back to the phase discriminator where its phase is compared with that of the input pulses. The voltage output V_1 from the discriminator is proportional to the phase difference $\theta_1 - \theta_2$

where

$$\theta_1 - \theta_2 = \int_0^T 2\pi(f_1 - f_2)\mathrm{d}t \tag{10.1}$$

since phase is the integral of frequency. From Fig. 10.1 we have

$$V_1 = \beta_1 \int_0^T 2\pi(f_1 - f_2)\mathrm{d}t \tag{10.2}$$

Remember that we have not yet specified f_1 and f_2: they will be assumed to be functions of time, either sinusoidal or transient. Now the perturbing frequency f_2 of the voltage appearing at the output of the line-scan oscillator is proportional to the voltage V_2 at its input, thus

$$2\pi f_2 = \beta_2 V_2 \tag{10.3}$$

The 2π is needed because β_2 is given in radians/second per volt rather than in hertz per volt.

Now, if we let V_1 be the amplitude of a steady-state cisoidal oscillation of angular frequency α and we let

$$p = j\alpha \tag{10.4}$$

we can calculate V_2 from an inspection of the circuit between the points in Fig. 10.1 where V_1 and V_2 are indicated. Thus

$$V_2[p] = \alpha_1 \frac{\alpha_2 + p}{p^2 + (\alpha_1 + \alpha_2 + \alpha_1\alpha_2/\alpha_3)p + \alpha_1\alpha_2} V_1[p] \qquad (10.5)$$

where

$$\alpha_1 = 1/R_1C_1 = 0.227 \text{ krad/s}$$
$$\alpha_2 = 1/R_2C_2 = 0.200 \text{ krad/s}$$
$$\alpha_3 = 1/R_1C_2 = 0.0091 \text{ krad/s}$$

We now replace V_1 by its value given in Equation 10.2 and V_2 by its value from Equation 10.3. First, let us work in angular frequencies

$$\omega_2 = 2\pi f_2 \quad \omega_2[p] = 2\pi f_2[p]$$

and

$$\omega_1 = 2\pi f_1 \quad \omega_1[p] = 2\pi f_1[p]$$

both functions of time, and therefore, spectrally, functions of complex (angular) frequency, $p = j\alpha$

Equation 10.2 may be written

$$V_1(t) = \beta_1[\int_0^T \omega_1(t)\mathrm{d}t - \int_0^T \omega_2(t)\mathrm{d}t$$

By rule (i), Equation 9.2 of Chapter 9, we may write

$$V_1[p] = \beta_1(\omega_1[p] - \omega_2[p])/p \qquad (10.2a)$$

and

$$V_2[p] = \omega_2[p]/\beta_2 \qquad (10.3a)$$

If we substitute for V_1 and V_2 their values from Equations 10.2a and 10.3a and put the results into Equation 10.5 we have

$$\omega_2[p] = \alpha_1\beta_1\beta_2 \frac{\alpha_2 + p}{p^3 + (\alpha_1 + \alpha_2 + \alpha_1\alpha_2/\alpha_3)p^2 + \alpha_1(\alpha_2 + \beta_1\beta_2)p + \alpha_1\alpha_2\beta_1\beta_2} \omega_1[p] \qquad (10.6)$$

where $\beta_1\beta_2 = 17.5$ krad/s per rad

Now

$$\omega_2(t) = \frac{\mathrm{d}}{\mathrm{d}t}\theta_2(t)$$

and

$$\omega_1(t) = \frac{\mathrm{d}}{\mathrm{d}t}\theta_1(t)$$

frequency (angular) being the derivative with respect to time of the phase. We have from rule (i), Equation 9.1,

$$\omega_2[p] = p\theta_2[p]$$

and

$$\omega_1[p] = p\theta_1[p]$$

Thus, Equation 10.6 is unaltered if we write it in terms of angular frequencies ω_2 and ω_1 or radians of phase θ_2 and θ_1. We shall, therefore, re-write it in terms of response R and excitation E, where R refers either to ω_2 or θ_2 and E refers either to ω_1 or θ_1. After insertion of the numerical values for the various constants and factorization—no simple task, we have

$$\mathrm{R}[p]/\mathrm{E}[p] = 3.97 \frac{0.2 + p}{(p + 0.367)(p + 0.473)(p + 4.58)} \tag{10.6a}$$

This is the transfer function of the complete circuit shown in Fig. 10.1. If we let $p = j\alpha = j2\pi f$, (Equation 10.4), f being the perturbing frequency of either the phase θ_1 or the frequency $\omega_1/2\pi$ of the input line-sync. pulses, then Equation 10.6a will, by suitable manipulation, give us the amplitude $|R|$ and the phase ϕ of the phase θ_2 or the frequency $\omega_2/2\pi$ of the output line-scan waveform oscillating sinusoidally around the frequency $\omega_2/2\pi$ at the frequency $f = \alpha/2\pi$; tricky, isn't it! Fig. 10.2 shows the modulus $|R/E| = |R|/|E|$ and the phase ϕ of the transfer function. At low frequencies, f, the flywheel circuit is capable of transmitting almost without attenuation, $|R/E| = 1$, the perturbation of the input phase θ_1 or frequency f_1 to the output phase θ_2 or frequency f_2, thus allowing the receiver scanning circuits to respond, or rather to follow, without error. At higher frequencies of perturbation the flywheel cannot follow, and at $f = 1000$ Hz, for example, the output perturbation magnitude is only one tenth of that at the input, thus leaving a visible picture perturbation having a value of nine tenths of that at the input. At 50 Hz the circuit magnifies the input perturbation by about 1.16, thus creating a visible disturbance equal to 16% of that at the input. The meaning of the phase angle $\phi < 0$ is that the output disturbance will occur at a time $\tau = \phi/\alpha$ after the input disturbance and therefore ϕ must be taken into account when subtracting the output perturbation from the input perturbation to find the actual error which will be visible to the television viewer. The reader who is interested in this problem must refer to Reference 13. Our purpose is to use this problem as an example of use of the Fourier integral and of convolution.

Let us now turn to the transient performance of the flywheel circuit. The transfer function, Equation 10.6a, is the Fourier transform of the time response of that circuit to a delta-function excitation, as remarked in Section 8.3; so in order to find R(t) when E(t) = $\delta(t)$ we must find the Fourier transform of the right-hand side of Equation 10.6a. Once again we rely upon Reference 8. Now

$$\frac{0.2}{(p + 0.367)(p + 0.473)(p + 4.58)} \doteqdot 0.2 \times (\text{coefficient } G(g) \text{ of pair } 452)$$

$$\frac{p}{(p + 0.367)(p + 0.473)(p + 4.58)} \doteqdot (\text{coefficient } G(g) \text{ of pair } 453)$$

Thus, using the asterisk notation to indicate response to delta-function excitation,

$$R^*(g) = 3.97 \{0.2 [4.11e^{-0.367g} - 4.21e^{-0.473g} + 0.11e^{-4.58g}]/1.83$$
$$-[1.51e^{-0.367g} - 1.99e^{-0.473g} + 0.48e^{-4.58g}]/1.83\}$$

or

$$R^*(g) = -1.5e^{-0.367g} + 2.5e^{-0.473g} - 1.0e^{-4.58g} \tag{10.7}$$

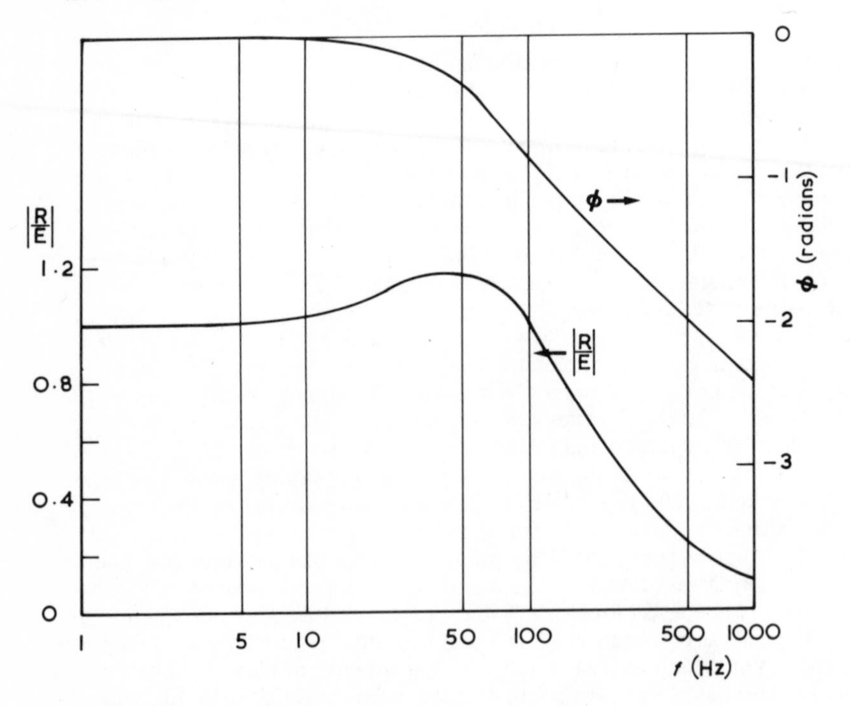

Fig. 10.2 The amplitude/frequency and the phase/frequency characteristics of the line-scan flywheel circuit

Of course, we may substitute t for g if we wish and either must be in units of milliseconds to ensure consistency with the constants such as 0.367, 0.473, etc. which are in units of kiloradians/second. Figure 10.3 shows $R^*(t)$. Now the actual disturbance to the television image will consist of a horizontal displacement having a magnitude equal to

$R^*(t) - E^*(t)$ where $E^*(t) = \delta(t)$. If $E^*(t)$ is a delta-function of phase then $R^*(t)$ will be a disturbance of phase and if $E^*(t)$ is a delta function of frequency, then $R^*(t)$ will be a disturbance of frequency. In either case, the excitation, being a delta function, will have come and gone at time zero, so Fig. 10.3, which shows $R^*(t)$ only, may be taken

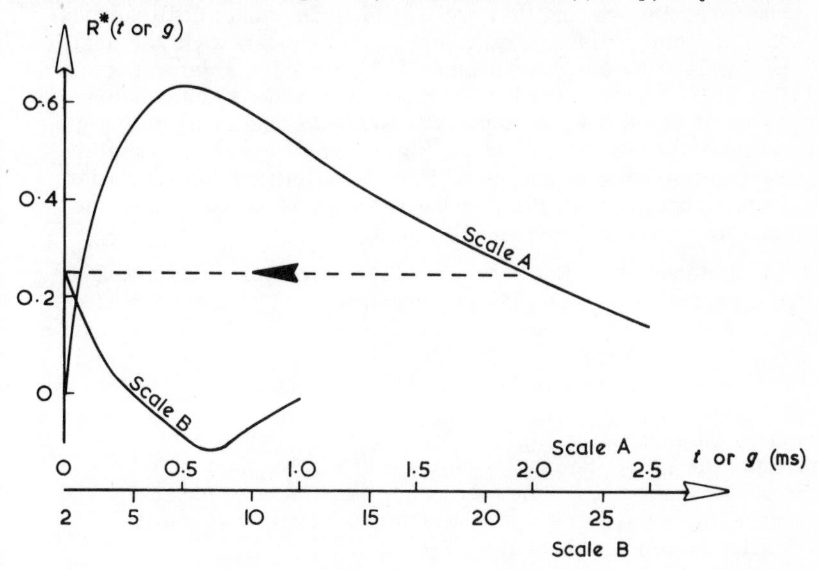

Fig. 10.3 Response of flywheel circuit to delta function excitation of phase or of frequency

as the picture disturbance $R^*(t) - E^*(t)$. If $E^*(t)$ is a delta function of phase we may evaluate the horizontal disturbance to the television picture as follows:

(a) Assume $\int_{-\infty}^{\infty} E^*(t)dt = U$ radian × milliseconds (or kiloradian seconds). This is the value or magnitude (not the peak or maximum) of the assumed delta function.

(b) Note that 2π radians of phase at the line-scan frequency of 15 625 Hz corresponds to the period of that frequency or 64 μs, whence one radian occupies a time period of $64/2\pi = 10.2$ μs.

(c) Note that, by definition, one horizontal (along-the-line) picture element has a time duration of $1/2f_c$ where f_c is the cut-off or highest frequency in the video bandwidth, which for UK 625-line television, is 5.5 MHz. Thus, one picture element occupies 0.091 μs and since there are only 52 μs available in each scanning line for picture information (the remaining 12 μs are taken up by synchronising signals in what is termed the 'line blanking interval') we find that the number of picture elements per line is 52/0.091 = 570.

(d) Note from (b) and (c) that there are 10.2/0.091 = 112 picture elements in one radian of phase of the line-scan waveform.

From the foregoing we find that one unit of U (in radian milliseconds) is worth 112 picture-element milliseconds, so we can scale the ordinates in Fig. 10.3 by multiplication by 112. Thus, for a delta function excitation of $U = 1$ radian millisecond we have a horizontal picture disturbance of $112 \times 0.63 = 70$ picture elements occurring at 0.5 ms after the epoch of the delta function, Fig. 10.3. Of course, the disturbance depends directly on the value U of the delta function excitation. The foregoing may seem rather obscure, so let us look at Equations 10.6a and 10.7 from a dimensional point of view.

First, the dimensions of Equation 10.6a can be examined by noting that the constant factor 3.97 is the product $\alpha_1\beta_1\beta_2$ whose dimensions are

$$[\mathrm{T}^{-1}]\ (\text{volts})\ [\mathrm{T}^{-1}\ (\text{volts})^{-1}] = \mathrm{T}^{-2}$$

where T is the dimension 'time'. (See Fig. 10.1 and the constants below Equation 10.5). Secondly, the term 0.2 in the numerator of Equation 10.6a is α_2 which has the same dimension as p, namely T^{-1}. Thirdly, the dimension of the denominator of Equation 10.6a is T^{-3}; thus finally, the dimension of $\mathrm{R}[p]/\mathrm{E}[p]$ is

$$\mathrm{T}^{-2}\,\frac{\mathrm{T}^{-1}}{\mathrm{T}^{-3}} = \text{a dimensionless number}$$

Now turn to Equation 10.7. What are its dimensions? To answer this we must turn to the Fourier transforms which are given as $\mathrm{G}(g)$ in pairs 452 and 453 of Reference 8. From inspection of these pairs it is evident that the dimension of the transform of

$$\frac{1}{(p + 0.367)\,(p + 0.473)\,(p + 4.58)}$$

is T^2, and the dimension of the transform of

$$\frac{p}{(p + 0.367)\,(p + 0.473)\,(p + 4.58)}$$

is T, thus the dimension of Equation 10.7 can be ascertained by replacing in the first (un-numbered) expression of Equation 10.7 the various factors and terms by their dimensions as follows

$$\begin{aligned}\text{Dimension of } \mathrm{R}^*(t) &= \mathrm{T}^{-2}\,\{\mathrm{T}^{-1}\,[\mathrm{T}^2] - [\mathrm{T}]\}\\ &= \mathrm{T}^{-1}\end{aligned}$$

The dimensions T^2 and T within the square brackets include the common denominator 1.83 which, by itself, has the dimension T^{-3}.

Now, since $R^*(t)$ has the dimension T^{-1} and it is the response to the delta function $E^*(t)$ assumed to have unit value, we see why we must re-insert $E^*(t)$ before we can actually evaluate $R^*(t)$. When an actual value of $R^*(t)$ is required, therefore, we have to assume a value for $E^*(t)$ and that value was postulated as U measured in radian milliseconds. So, for a practical application we should re-write Equation 10.7 as

$$R^*(t) = (-1.5e^{-0.367t} + 2.5e^{-0.473t} - e^{-4.58t})U \tag{10.7a}$$

Since the dimension of U is T or, if you like, θT where θ represents radians (a dimensionless quantity) we obtain $R^*(t)$ as a dimensionless quantity, or if you prefer, as a number of radians. Some readers may find this explanation unnecessary, but the practical use of the delta function is often avoided and I believe this is sometimes due to lack of clear understanding of the dimensions of that function in a real practical case.

Let us now examine the response $R_1(t)$ to an excitation consisting of a unit-step of phase, $E_1(t)$. Since the unit-step is the integral of the unit-impulse or delta function we know that, for linear circuits, the response to unit-step will be the integral of the response to unit-impulse; so

$$R_1(t) = \int_0^t R^*(g)dg \tag{10.8}$$

and from Equation 10.7

$$R_1(t) = \int_0^t [-1.5e^{-0.367g} + 2.5e^{-0.473g} - e^{-4.58g}]dg$$

$$= 1 + 4.08e^{-0.367t} + 0.22e^{-4.58t} - 5.3e^{-0.473t} \tag{10.9}$$

The actual picture disturbance in radians, due to a unit-step excitation of one radian, will be

$$R_1(t) - E_1(t) = R_1(t) - 1$$

or, from Equation 10.9,

$$R_1(t) - E_1(t) = 4.08e^{-0.367t} + 0.22e^{-4.58t} - 5.3e^{-0.473t} \tag{10.10}$$

This function is shown in Fig. 10.4. It can be seen that the picture disturbance is equal to the full value of the disturbing excitation at $t = 0$, but that it diminishes to zero after about 15 milliseconds or the duration of about 14 000/64 = 22 lines of scanned picture. The initial value of the picture disturbance, indicated as —1 in Fig. 10. 4, represents radians of picture shift per radian of disturbing excitation, or microseconds of picture shift per disturbing microsecond, or picture elements of picture shift per picture element of disturbance, or hertz

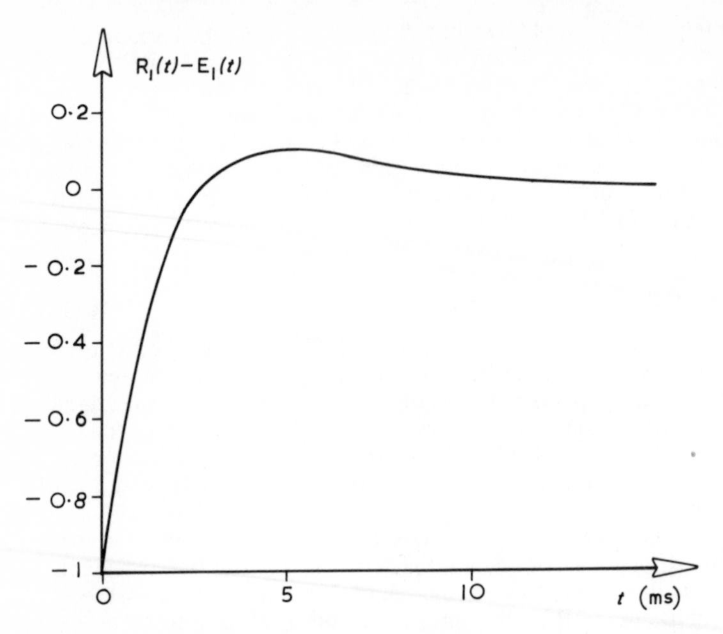

Fig. 10. 4 Response in microseconds of flywheel circuit, to a microsecond unit-step excitation

of line-scan frequency per hertz of synchronising pulse frequency, etc.

The above example was rather lengthy and involved, but it is a quite practical case, which is in process of being converted into a code of practice specifying the maximum permissible disturbance to the timing of the flow of synchronising pulses in a television signal.

10. 2 ELEMENTARY APERTURE THEORY

10. 2. 1 A Gaussian lens

The simplest possible aperture that we can study has only one dimension, but the extension of the theory of aperture distortion, or rather, the distortion caused by an aperture, to two dimensional apertures involves no further philosophical difficulties, although the mathematics become more complicated.

Consider an optical lens being used to focus an image of a scene on to an opaque screen in the image plane of the lens. Let us consider a simple element of the scene which happens to lie on the principal axis of the lens, that is, its image will lie in the very centre of the image plane of the lens. The reason for making this stipulation is simply to ensure that any symmetry pertaining to the element in the scene is retained in its image—a lens normally has no geometric distortion to objects lying within the region of the 'paraxial' (parallel and central) rays. We now make the further simplifying assumption that the picture element in the object scene has only one dimension, say in the x direction. Of course, this is never true, but it enables calculations to be made about the effect of lens aperture distortion in the horizontal direction in television images that are always built up from horizontal scanning lines. We shall consider, later, the vertically dimensioned aperture formed by a television scanning line having a finite, rather than zero, width.

We now assume that the element in question consists of a delta function of brightness as a function of horizontal distance in the x direction only. Will the horizontal profile or space function of the image formed by the lens also take the form of a delta function? No, it won't, because the lens cannot be 'ideal' in the 'filter' sense. The lens may be considered in Fourier or spectral terms as a bandpass filter, the frequencies of the spectral components of the image formed by the lens being measured not in hertz or cycles per second, but in cycles per millimetre. Thus, the reader can appreciate that there is a limit to the ability of the lens to respond to very fine patterns of vertical black-and-white lines (horizontal resolution), just as there is a limit to the field-of-view of the lens and therefore to its ability to reproduce very coarse patterns of vertical black-and-white lines. If the reader is sceptical about the last phrase above, just image a set of infinitely tall black-and-white bars, each bar being a kilometre wide! All lenses are, therefore, bandpass filters capable of reproducing patterns of black-and-white lines that are neither infinitely coarse, (frequency: zero cycles per millimetre), nor infinitely fine, (frequency: infinity cycles per millimetre). Incidentally, frequencies measured in cycles per unit distance, such as the millimetre, are termed spatial frequencies. Into what kind of a profile or space-versus-distance function will the lens transform our delta function test element? The answer depends upon the lens; a $\cos^4 x$ function taken between $-\pi/2$ and $+\pi/2$ is one possibility. This form has the disadvantage of limiting the smearing or defocusing effect of the lens to $\mp\pi/2$, which is clearly untrue, although adequate for many purposes. We shall assume that the space function response of the lens to a delta space function is Gaussian, thus

$$\mathrm{R}_0(x) = B_0 \mathrm{e}^{-x^2/2X^2} \tag{10.11}$$

where the delta function excitation has an area A measured in nit × millimetres, B_0 is the brightness or luminance in nits measured at the centre of the image ($x = 0$), and X is the standard deviation of the Gaussian function. The nit is the unit of luminance. It is equal to 10.8 footlamberts or 3.1 candles per square metre. The coefficient B_0 depends upon both the lens and the value or area A of the delta function excitation, whilst X is a lens constant.

Now let us ask ourselves what will be the form of the horizontal space function response of the lens to a black-to-white transition in the scene, the transition being horizontal; that is, the left-hand half of the scene being black and the right-hand half white: in other words, a unit-step transition. We have to convolve $R_0(x)$ with $h(y)$, if $h(y)$ expresses the unit-step transition from black to white. We thus require

$$R_1(x) = R_0(x)*h(y) \tag{10.12}$$

This is similar to the convolution of a parabola with a Gaussian function as dealt with in Section 4.2.3 except that the unit-step is not a continuous function between $-\infty$ and ∞ as was the parabola. The upper limit of the integral implied by Equation 10.12 must be restricted to $u = x + y$, see Fig. 10.5; thus, letting $y = u - x$, we have

$$R_1(u) = \int_{\infty}^{u} R_0(x)h(u - x)dx \tag{10.13}$$

Since the integral is limited to u, (in the positive direction of x), $u - x \geqslant 0$, and so whatever the value of $-\infty < u < \infty$ we never have to deal

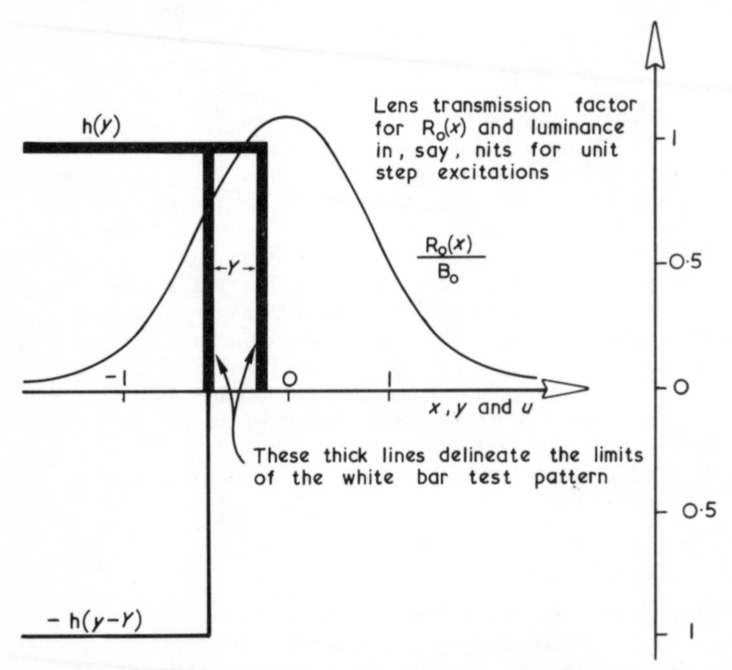

Fig. 10.5 Response of an optical lens to a delta space function of illumination and graph of a rectangular space function excitation

with a negative value of the variable of the function h, so we do not violate the condition that a unit step exists only for positive values of its independent variable. Since $u - x \geqslant 0$, $h(u - x) = 1$, so

$$R_1(u) = \int_{-\infty}^{u} R_0(x)dx \tag{10.14}$$

Thus, the lens, as an image-forming device, being linear—twice the excitation produces twice the output—we find, once again, that the response to a unit step is the integral of the response to a delta function. From Equations 10.11 and 10.14

$$R_1(u) = B_0 \int_{-\infty}^{u} e^{-x^2/2X^2} dx \tag{10.15}$$

Now

$$\int_{-\infty}^{u} = \int_{-\infty}^{\infty} - \int_{u}^{\infty} = \int_{-\infty}^{\infty} - [\int_{0}^{\infty} - \int_{0}^{u}]$$

but the integrand is an even function, so $\int_{-\infty}^{\infty} = 2\int_{0}^{\infty}$ and finally

$$\int_{-\infty}^{u} = \int_{0}^{\infty} + \int_{0}^{u}$$

Now

$$B_0 \int_{0}^{\infty} e^{-x^2/2X^2} dx = \sqrt{(\pi/2)} B_0 X \tag{10.16}$$

(see integral no. 492, p. 63, in Reference 5) and

$$B_0 \int_{0}^{u} e^{-x^2/2X^2} dx = X\sqrt{2} B_0 \int_{0}^{u/X\sqrt{2}} e^{-t^2} dt$$

after changing the variable by letting $x^2/2X^2 = t^2$

The function $\int_{0}^{t} e^{-t^2} dt$ is often written $(1/2)!\ E_2(t)$ or $(1/2)!\ \mathrm{erf}t$, (erf for error function) (see Reference 5, page 116 and Reference 15). Now $\frac{1}{2}! = \Gamma(3/2)$ and the gamma function of 3/2 is $\sqrt{\pi}/2$; thus

$$X\sqrt{2}\, B_0 \int_{0}^{u/X\sqrt{2}} e^{-t^2}\, dt = XB_0\sqrt{(\pi/2)}\ \mathrm{erf}\ (u/X\sqrt{2})$$

Finally

$$R_1(u) = XB_0\sqrt{(\pi/2)}\ [1 + \mathrm{erf}\ (u/X\sqrt{2})] \tag{10.17}$$

with $-\infty < u < \infty$.

Now what is $\mathrm{erf}(u/X\sqrt{2})$ when $u < 0$? It is easy to prove that $\mathrm{erf}(-t) = -\mathrm{erf}(t)$. Consider the series expansion of $\int e^{-t^2}dt$, thus

$$\int e^{-t^2}dt = \int [1 - t^2 + \frac{t^4}{2!} - \frac{t^6}{3!} + \frac{t^8}{4!} - \ldots]\,dt$$

or

$$(\sqrt{\pi}/2)\mathrm{erf}(t) = t - \frac{t^3}{3} + \frac{t^5}{2!5} - \frac{t^7}{3!7} + \frac{t^9}{4!9} - \ldots$$

Now substitute $-t$ for t

$$(\sqrt{\pi}/2)\mathrm{erf}(-t) = -t + \frac{t^3}{3} - \frac{t^5}{2!5} + \frac{t^7}{3!7} - \frac{t^9}{4!9} + \ldots$$

$$= -1(\sqrt{\pi}/2)\mathrm{erf}(t)$$

which is what we set out to prove. In fact, $\mathrm{erf}(t)$ varies from -1 to $+1$ as t varies from $-\infty$ to $+\infty$. We could now plot Equation 10.17 as a graph with $-\infty < u < \infty$ as variable, but it may be more instructive if we now change our optical test pattern from a simple black-to-white transition to a vertical white bar on an otherwise black background. For this purpose we change the optical excitation function from being a single unit step $h(y)$, to the sum of the original unit step $h(y)$ followed by a negative unit step a distance Y behind the first; that is $-h(y - Y)$, Fig. 10.5. The response to this combination of unit steps will consist of the original response [to $h(y)$] given by Equation 10.17 plus the response to the negative, 'delayed' unit step which, from Equation 10.17, adapted, is

$$-R_1(u - Y) = -XB_0\sqrt{(\pi/2)}\,\{1 + \mathrm{erf}\,[(u - Y)/X\sqrt{2}]\}$$

So, the final response, $R_2(u)$, to the white bar test pattern will be

$$R_2(u) = R_1(u) - R_1(u - Y) = XB_0\sqrt{(\pi/2)}\,[\mathrm{erf}(u/X\sqrt{2}) - \mathrm{erf}(\overline{u - Y}/X\sqrt{2})] \qquad (10.18)$$

We can easily make Equation 10.18 look more symmetrical by letting

$$u/X\sqrt{2} = s + w/2$$

and

$$u/X\sqrt{2} - Y/X\sqrt{2} = s - w/2$$

whence the response $R_2(s)$ can be written

$$R_2(s) = XB_0\sqrt{(\pi/2)}\,[\mathrm{erf}(s + w/2) - \mathrm{erf}(s + w/2)] \qquad (10.18a)$$

with

$$s = u/X\sqrt{2} - Y/2X\sqrt{2}$$

$$w = Y/X\sqrt{2}$$

Equation 10.18a is symmetrical in s, but Equation 10.18 is not symmetrical in terms of u, whereas Equation 10.17 was (skew) symmetrical in terms of u. The reason for the lack of symmetry of Equation 10.18 in terms of y is because our white bar itself was constructed in an asymmetrical way, being the addition of a normal unit step $h(y)$ and a negative unit step $-h(y - Y)$. In order to obtain symmetry in terms of u we should have had to 'advance' both $h(y)$, by $+Y/2$, making $h(y + Y/2)$, and $-h(y - Y)$, by $+Y/2$ making it $-h(y - Y/2)$. The lack of symmetry is, however, of no importance. Fig. 10.6 shows three examples of $R_2(s)/XB_0\sqrt{(\pi/2)}$; when $Y = X/\sqrt{2}$, $Y = X\sqrt{2}$, $Y = 2X\sqrt{2}$. We could have normalised the excitations that gave rise to the three curves in Fig. 10.6 by increasing the magnitudes of the two unit steps in Fig. 10.5 in inverse proportion to the gap Y between them, so as to

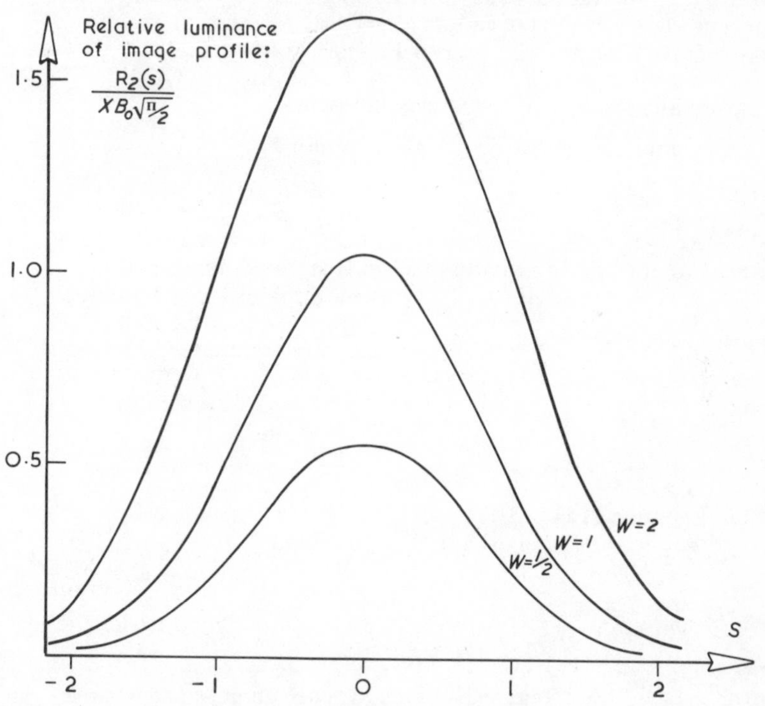

Fig 10.6 Response of Gaussian-aperture lens to rectangular-profile excitation

maintain constant the area of the rectangular pulse formed by them. This would be similar to the 'dodge' mentioned at the end of Section 4.2.3 where it was suggested that a delta function can be constructed from a rectangle of width a and height $1/a$, and then letting $a \to 0$; although we are not, here, attempting the construction of delta functions. Note that keeping constant the area of the rectangular pulse, which is the *profile function* of the white bar, would not equalise the maximum values of the three responses, because width of rectangle is not exactly compensated by height until the rectangle is so narrow that its spectrum, in terms of spatial frequencies, is uniform up to the highest spatial frequency contained in the passband of the lens, considered as a filter of spatial-frequency components. When that occurs, the rectangular profile has become, in practical terms, a delta function. Similarly, if the rectangular test pulse were made very wide indeed, the progressively diminishing response of the lens to very low spatial frequencies would begin to make itself felt in an asymptotic fashion. The above remarks relating to the spectral content of the white bar lead to the question 'what is the spectral content of the white bar and how does it depend upon the width of the bar?' (the spectrum of a rectangular pulse is a function that appears again and again in physics and engineering) and the inverse question 'what is the space or time time function whose spectrum is a rectangle?'. To answer these two questions requires the solution of two Fourier transforms.

10.2.2 Spectral response of a rectangular pulse

Consider a rectangular pulse or profile of width Y

$$\mathrm{P}(x) = 1 \quad -Y/2 < x < +Y/2$$

$$\mathrm{P}(x) = 0 \quad |x| > Y/2$$

Its spectrum, according to Equation 6.2 in Chapter 6, is

$$\mathrm{P}[z] = \int_{-\infty}^{\infty} \mathrm{P}(x)\mathrm{e}^{-j2\pi zx}\mathrm{d}x$$

$$= \int_{-Y/2}^{Y/2} \mathrm{e}^{-j2\pi zx}\mathrm{d}x$$

$$= \int_{-Y/2}^{Y/2} \cos 2\pi zx \cdot \mathrm{d}x$$

the term

$$-j\int_{-Y/2}^{Y/2} \sin 2\pi zx \cdot \mathrm{d}x$$

being zero, because the integrand is an odd function of x: the area to the left of $x = 0$ being the negative of the area to the right of that point. Finally

$$\mathrm{P}(z) = (1/\pi z) \sin \pi zY$$

or, for convenience

$$\mathrm{P}[z] = \frac{Y\sin\pi Yz}{\pi Yz} \qquad (10.19)$$

where z is the frequency in cycles per unit of x of each of the sine-waves that make up the spectrum $\mathrm{P}[z]$. Figure 10.7 shows $\mathrm{P}[z]/Y$

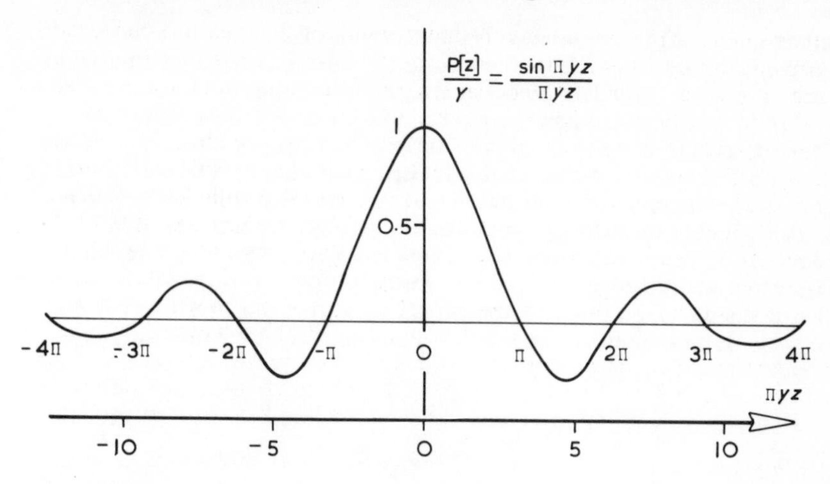

Fig. 10.7 The $(\sin x)/x$ or sinc function

plotted as a function of πYz. The first thing to note about this function, $(\sin \pi Yz)/\pi Yz$—sometimes called the 'sinc' function, sometimes the 'sampling' function—is that the maxima, other than the maximum maximorum that occurs at $\pi Yz = 0$, do not occur at exactly the values of πYz which maximise the function $\sin \pi Yz$. If we equate to zero the derivative with respect to x of $(\sin x)/x$ we obtain

$$\tan x = x$$

which gives maxima and minima of $(\sin x)/x$ as follows:

(a) first minimum of $(\sin x)/x$ occurs when $x = 4.49$, whereas minimum of $\sin x$ would occur when $x = 3\pi/2 = 4.72$

(b) second maximum (the first one occurs at $x = 0$) of $(\sin x)/x$ occurs when $x = 7.73$, whereas second maximum of $\sin x$ would occur at $x = 5\pi/2 = 7.86$

(c) second minimum of $(\sin x)/x$ occurs when $x = 11.086$, whereas second minimum of $\sin x$ would occur at $x = 7\pi/2 = 11.$

We see that as x increases, the maxima and minima of $\sin x/x$ approach the positions (along the x-axis) that would be occupied by the maxima and minima of $\sin x$ itself.

The second and more important feature of the $(\sin x)/x$ function is that all its zeroes are exactly π apart, each from the next, just as are the zeroes of $\sin x$.

The third feature is that, unlike $\sin x$, the function $(\sin x)/x$, although continuous for $-\infty < x < \infty$, has an envelope that diminishes according to $1/x$. The last two features will be discussed later.

Let us now return to the question regarding the effect of the width Y of the rectangle of height 1, as shown in Fig. 10.5 as the test pattern profile for the lens with the Gaussian response to a delta function of luminance. But first, we must remember that the pulse of light of width Y will usually be composed of incoherent light that, unless it comes from a laser source and is therefore coherent, can never be negative since it, in fact, is itself a beam of electromagnetic energy rather than a beam of force such as a coherent magnetic or electric force that would be the case if we were dealing with broadcasting signals or voltages applied to electric circuits, etc. We therefore have to transform the function shown in Fig. 10.7 in such a way that all the negative lobes of the $(\sin x)/x$ curve become positive; as if we had passed the function through a full-wave rectifier! For our lens calculations, therefore, the function $P[z]$ becomes

$$P_+[z] = |P[z]| = Y|\sin \pi Yz/\pi Yz|^* \tag{10.19a}$$

We now have the spectrum of the rectangular pulse and we must move on and calculate the response-versus-frequency characteristic of the lens in order to find out the ratio of pulse width Y to the width X of the spatial-response function of the lens such that the lens passband is adequately supplied with frequency components from the pulse spectrum. For pulse widths narrower than this special value of Y there will be adequate compensation for diminishing widths by proportional increases in heights.

The spectrum of Expression 10.19a, resulting as it does from an idealised rectangular profile, is not a physical reality, because its total power or energy is infinite; thus

$$\int_{-\infty}^{\infty} \left|\frac{\sin x}{x}\right| dx = 2\int_{-\infty}^{\infty} \left|\frac{\sin x}{x}\right| dx \to \infty$$

What happens is that in forming a rectangular profile by means of a rectangular slit in an otherwise opaque screen placed between the lens and the source of light, the edges of the slit would give rise to diffraction of the light thus modifying the shape of the edges of the profile and affecting the higher frequency components of the spectrum of the profile. Negligible error will result from calculations involving the lower frequency portion of the spectrum, but when the slit, and therefore the profile, is very narrow, diffraction will predominate and calculations of the type we are making here will become quite meaningless.

10.2.3 Spectral response of a Gaussian profile

We therefore calculate the 'passband' of the lens whose delta function response is given by Equation 10.11. Again, we calculate the Fourier transform of the response:

$$R_0[z] = B_0 \int_{-\infty}^{\infty} e^{-x^2/2X^2} \cdot e^{-j2\pi zx} dx \tag{10.20}$$

similar to Equation 6.2 in Section 10.2.2.

This transform is given in Reference 8 as Pair No. 705.1, but it is easy to simplify it to a well known infinite integral whose solution was given in Section 4.2.3. We perform the following operations on the integrand:

$$\exp(-x^2/2X^2) \cdot \exp(-j2\pi zx) = \exp\{-(x^2 + j4\ \pi X^2 zx)/2X^2\}$$

We now convert the exponent into a perfect square and subtract from that square its last term, thus

$$(a + b)^2 - b^2 = a^2 + 2ab$$

as we did to arrive at Equation 4.25 in Section 4.2.4. The integrand now becomes, remembering that $j^2 = -1$,

$$\exp\{-[(x + j2\pi x^2 z)^2 + 4\pi^2 x^4 z^2]/2x^2\}$$

$$= \exp(-2\pi x^2 z^2) \cdot \exp\{-(x + (j2\pi X^2 z)^2/2X^2\}$$

Now we let

$$r = (x + j2\pi X^2 z)/X\sqrt{2}$$

whence

$$x = rX\sqrt{2} - j2\pi X^2 z \text{ and } dx = X\sqrt{(2)}dr$$

and thus from Equation 10.20

$$R_0[z] = B_0 X\sqrt{2}\exp(-2\pi^2 x^2 z^2)\int_{-\infty}^{\infty} e^{-r^2}\,dr \qquad (10.21)$$

and, from Section 4.2.3

$$R_0[z] = B_0 X\sqrt{(2\pi)}\exp(-2\pi^2 X^2 z^2) \qquad (10.22)$$

$R_0[z]$ is, like $R_0(x)$, a Gaussian function, see Fig. 10.5. This is a remarkable property of that function—its Fourier transform is also Gaussian! It is not the only function having this property; the hyperbolic secant, sech$x$, also possesses it and there may be others for all I know.

Equation 10.22 represents the filtering characteristic of our lens regarded as a low-pass filter. We took no account of its inability to pass very low spatial frequencies. The full band pass characteristic would have a vertical 'crack' in the curve of $R_0[z]$ plotted as a function of z, just around the frequencies close to $|z| = 0$. We shall take no account of this aspect.

10.2.4 Effect of pulse width on lens response

Figure 10.6 shows that the response of the lens to the rectangular pulse decreases with decreasing pulse width, Y, ($W = Y/X\sqrt{2}$), but that the rate of decrease is less than linear: if we consider the maximum response $R_2(0)$ that occurs at the frequency $s = 0$, we find, from Fig. 10.6, the values in Table 10.1

Table 10.1

W	$R_2(0)/\{XB_0\surd(\pi/2)\}$	Values in 2nd column normalised to value at which $W = 1$
2	1.685	1.62
1	1.045	1
1/2	0.55	0.53

Thus, halving the pulse width from $Y = 2X\sqrt{2}$ to $Y = X\sqrt{2}$ multiplies $R_2(0)$ by 1/1.62 instead of by 1/2 as would be the case if the width of the pulse spectrum adequately exceeded the pass band of the lens response/frequency characteristic. If this condition were achieved, then, as stated earlier, a halving of pulse width would halve the response of the lens and this could be compensated by a doubling of the pulse magnitude. Linearity is, however, being approached as Y is reduced from $Y = X\sqrt{2}$, ($W = 1$), to $Y = X/\sqrt{2}$, ($W = 1/2$)—a further halving; for now we find that $R_2(0)$ is reduced to 0.53 of its value for $Y = X\sqrt{2}$ or $W = 1$. True linearity would make the reduction factor 0.5 instead of 0.53. How should we calculate the value of Y that makes the uniform portion of the pulse spectrum just exceed the lens's pass band? An exact calculation would be quite complicated, but an approximation may be reached by examining Fig. 10.7 to find at what frequency z the spectrum ceases to be uniform. Although Fig. 10.7 shows P[z] rather than $\mathrm{P}_+[z]$, this will not be significant, as will be obvious from what follows. The shape of P[z] is such that there is no uniform portion, so we have to make an arbitrary decision as to the meaning of the word 'uniform'. If we consider the frequency z at which $\pi Yz = 1$ we note that the spectrum has fallen by 16% from its value at $z = 0$. Broadcasting engineers would put this into decibels thus $10 \log_{10} 0.84 = -0.75$ dB, (P[z] or $\mathrm{P}_+[z]$ representing power or energy, not force or voltage, etc.). Thus quite arbitrarily, let us pretend that up to the frequency $z_1 = 1/\pi Y$ the spectrum of the rectangular profile is sufficiently constant to be taken to be uniform. This kind of decision is typical of those that engineers are required to make in the course of their work. We now ask 'how much of the total width of the response/frequency characteristic of the lens do we need to fill with "uniform" spectrum from the rectangular profile (with $|\sin z/z|$ spectrum) in order to ensure adequate compensation to take place between height and width Y of the rectangular profile?' Here, again, to avoid having to plot a number of curves and find the answer which, in any case, will not be a precise and categorical one, we shall assume that the pass band of the lens will be adequately filled with spectral components of the rectangular-pulse excitation if the frequency z_1 is sufficiently high to correspond with a low value of lens response. Thus, let z_2 be that frequency for which the response of the lens has fallen to, say, 1/20th of its maximum response, (at zero frequency).

$$R_0[z_2]/R_0[0] = 0.05$$

Equation 10.22 yields

$$z_2 = \pm 0.39/X$$

Now, with $z_1 = 1/\pi Y$, if we equate z_1 to z_2 we have

$$Y/X = 0.82$$

and

$$W = 0.58$$

So, finally, we can assume with reasonable confidence that if $w \leq 0.6$ the rectangular excitation will behave like a delta function and the response of the lens will remain unchanged for ever-decreasing values of Y provided that the area of the rectangular profile is maintained constant. This approximately confirms the conclusion we drew from Table 10.1 that the change from $W = 1$ to $W = 1/2$ was already beginning to enter the linear region of response $R_2(0)$ as a function of Y.

10.2.5 A numerical example

Here are some practical considerations enabling us to put figures into the results we have achieved. A certain zoom lens, which is a complicated optical device having more than a dozen air-to-glass interfaces, has a measured response factor of about one-half to a test pattern consisting of a set of black-and-white bars at a frequency of eight cycles per millimetre; the test pattern being placed on the optical axis of the lens which, itself, has been set to the 20-inch focal-length position. This spatial frequency of 8 cycles/mm corresponds approximately with the highest video frequency that television systems are designed to pass without undue attenuation. It may, therefore, be concluded that the optical lenses used in television cameras do make a significant contribution to the attenuation in the overall response/frequency characteristic of the system.

If, using Equation 10.22, we calculate X when the response has fallen to one-half at $z = 8$ c/mm, we may write

$$R_0[8]/R_0[0] = \exp[-2\pi^2 X^2 8^2)$$
$$= 0.5$$

or

$$X = 0.0256 \text{ mm}$$

Now the width of one of the white bars in the test pattern having a spatial frequency of 8 mm will be $^1/_{16}$ mm; that is,

$$Y = 0.0625 \text{ mm}$$

and this may be taken as representing the width of the smallest single picture element that the television system is designed to reproduce.

We thus have, from immediately below Equation 10.18a

$$W = 0.0625/0.0256\sqrt{2}$$
$$= 1.73$$

So the way in which the lens would reproduce the width of the white bar of width $Y = 0.0625$ mm can be judged by an inspection of the two curves for $W = 1$ and $W = 2$ in Fig. 10.6. Equation 10.18 is shown plotted as a function of u in Fig. 10.8, taking Y and X as given above. The maximum value of $R_0(u)/XB_0\sqrt{(\pi/2)}$, namely 1.555 (Fig. 10.8), is

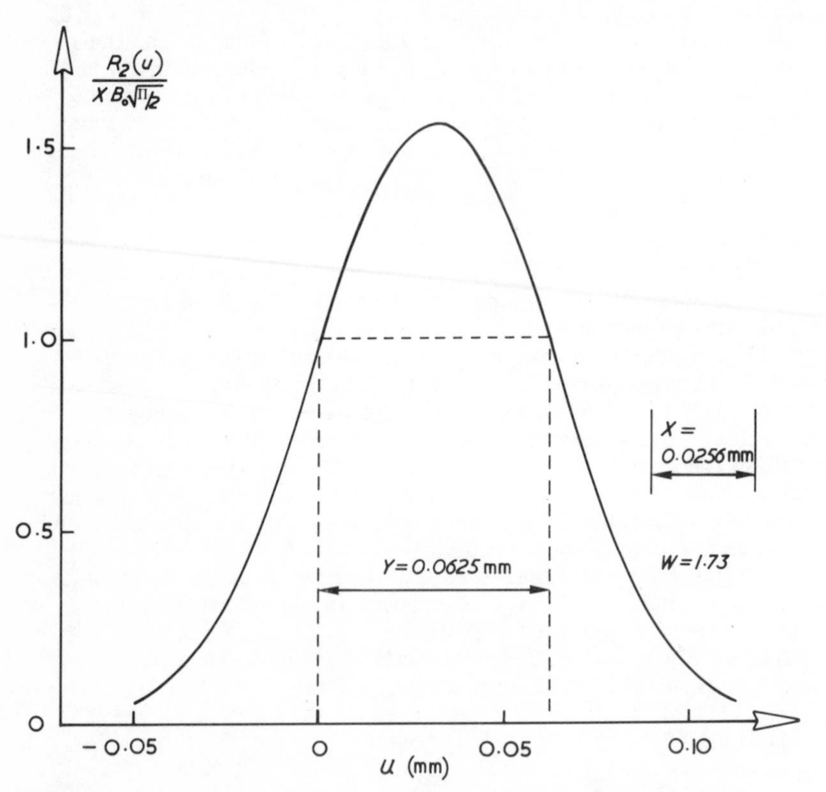

Fig. 10.8 Response of a zoom lens to a white-bar test pattern

quite arbitrary, because we do not know the value of B_0; but we can appreciate the way in which the rectangular profile of the white bar has been broadened and the vertical edges distorted by passage through the lens. The lack of symmetry around the ordinate at $u = 0$ has been mentioned earlier as being due to the asymmetrical method of construction of the rectangular pulse itself, this being shown by the dashed-line rectangle in Fig. 10.8.

10.3 RESPONSES OF AN IDEAL LOW-PASS FILTER

Having given some thought to the optical lens as, on the one hand an aperture and on the other hand as a filter of spatial frequencies, let us consider a straightforward electrical problem; namely, the performance of an ideal low-pass filter. Figure 7.2 showed a very simple low-pass filter—the series R, shunt C circuit. Its principal property is to apply increasing attenuation to sinusoidal voltage excitations of increasing frequency. In fact, it offers no attenuation to a voltage at d.c. and infinite attenuation to a voltage having infinite frequency. An idealised low-pass filter would be one with a rectangular passband, passing without attenuation voltages having frequencies between zero and some value F. The phase difference between output and input voltages must be a linear function of frequency in order to avoid any distortion other than a constant time delay. Figure 10.9 shows such characteristics. Note that I have drawn the pass region in such a way as to include negative frequencies as well as positive ones. One should always do this, because it is impossible to distinguish between a negative frequency and a positive one and the filter is in a similar state of ignorance. Thus, a bandpass filter's response/frequency characteristic would be drawn to show the rectangle of Fig. 10.9 moved entirely to the right of the zero-frequency ordinate and there would be a mirror image of it entirely to the left of the zero-frequency ordinate as well. Of course, the bandwidth of such a filter would be $2F$ rather than F.

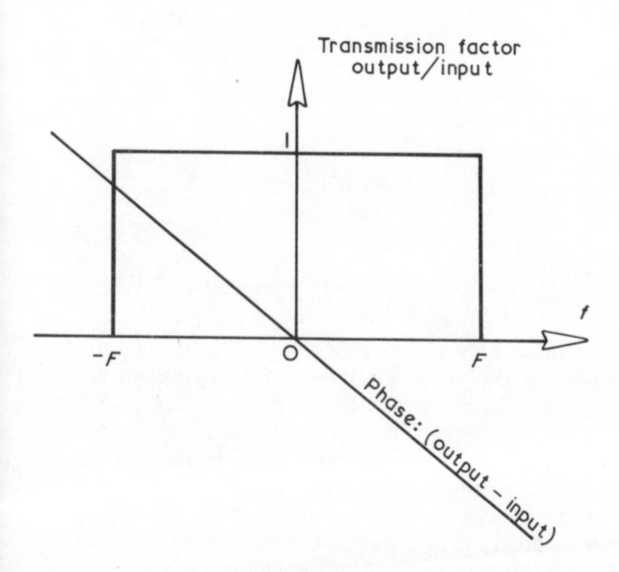

Fig. 10.9 Characteristics of an ideal low-pass filter

10.3.1 To a delta function excitation

The spectrum of a delta function of time is uniform at all frequencies and if the area or value of it is, say, A volt-seconds, then its spectrum, following Equation 7.2 of Chapter 7, will be simply A.

Now the transfer function of the ideal filter shown in Fig. 10.9 will be e^{-jaf} if $-a$ is the slope of the phase/frequency characteristic, shown as linear in the figure. The frequency limits over which this assumption is valid are, of course, $\pm F$. This transfer function can be seen to be correct if we expand it as follows:

$$e^{-jaf} = \cos af - j\sin af$$

and the phase, ϕ, of such a function is

$$\begin{aligned}\phi &= \arctan\left(-\frac{\sin af}{\cos af}\right)\\ &= \arctan(-\tan af)\\ &= -af\end{aligned}$$

That such a transfer function (modulus one, and phase linear with frequency) introduces no distortion except a time delay will be seen by what follows.

We now use Equation 6.6 in Chapter 6, to take the Fourier transform of the complete spectrum of the output from the filter.

$$R_0(t) = \int_{-F}^{F} Ae^{-jaf} \cdot e^{j2\pi tf}\,df \qquad (10.23)$$

$$= A\int_{-F}^{F} e^{j(2\pi t-a)f}\,df$$

$$= \frac{A}{j2\pi(t-a/2\pi)}\left| e^{j(2\pi t-a)f}\right|_{-F}^{F}$$

$$R_0(t) = 2AF\,\frac{\sin 2\pi(t-a/2\pi)F}{2\pi(t-a/2\pi)F} \qquad (10.24)$$

The argument or angle of the sine wave is $2\pi F\,(t - a/2\pi)$ instead of simply $2\pi Ft$, and the mathematics requires careful interpretation in this case. The centre of gravity or mean value of the function $(\sin x)/x$ occurs when $x = 0$, and what our ideal filter has done is to delay the arrival of the centre of gravity of the output for a time equal to $a/2\pi$; however, the function $(\sin x)/x$, being symmetrical around $x = 0$, can be said to start at $x \to -\infty$ and end when $x \to \infty$, but nothing can emerge from the filter until $a/2\pi$ seconds after the application of the excitation. The way out of this paradox is to conclude that the transmission time through an ideal filter is $a/2\pi = \infty$ and this is why the ideal filter is usually treated as a phaseless device. The time delay of a single section of the most elementary low-pass filter, Fig. 7. 2, of Section 7. 1, the *RC* circuit, is about $1/2f_C$, where f_C is the nominal cut-off frequency at which the attenuation is 3 dB $(1/\sqrt{2})$. Two such filters connected in tandem, but separated by an isolating, one-way device such as a transistor, double the time delay, but sharpen the shape of the cut-off region, and thus the price paid for the achievement of an infinitely sharp cut-off such as is assumed for the ideal filter is an infinite time delay.

The time delay $a/2\pi$ of any linear circuit can be calculated from the slope of the phase/frequency characteristic; thus, since $\phi = -af$ we have $a = -\phi/f$. If, in the general case, ϕ is not a linear function of f we have to consider the ratio $a = \phi/f$ over a small region df for which linearity may be assumed. Thus, finally, the time delay for a group of frequency components of width df may be written

$$\tau = -(1/2\pi)\mathrm{d}\phi/\mathrm{d}f \tag{10.25}$$

$\mathrm{d}\phi/\mathrm{d}f$ being always negative, this makes τ positive, because we have called it a 'delay'. The $1/2\pi$ factor converts the phase ϕ, which will be in radians, into revolutions or cycles which is what the time τ 'is interested in'.

Very often, the ideal low-pass filter, Fig. 10. 9, is used without any assumptions about its phase characteristics being involved, as mentioned above, but it is better always to bear in mind the fact that an amplitude/frequency characteristic without an accompanying phase/frequency characteristic is an impossibility, since, at very best, an electric signal or a disturbance of any kind must take time to traverse a circuit or a medium. Only the ether and a loss-free transmission line terminated by its characteristic impedance transmit without distortion, although the ether requires a time delay of one second for every 300 000 kilometres travelled and the transmission line requires a time delay of $x\sqrt{(LC)}$ where x is the length of the line and L and C are the inductance and capacitance per unit length, see Section 8. 5. 2 of Chapter 8 and Equation 8. 45. Equation 8. 44 shows that the phase/frequency characteristic of the terminated, loss-free line is $\phi = -\omega x\sqrt{(LC)}$ and, since $\omega = 2\pi f$

$$-(1/2\pi)\mathrm{d}\phi/\mathrm{d}f = x\sqrt{(LC)}$$

$$= \tau, \text{ the time delay!}$$

The phase/frequency characteristic of the ether can readily be obtained thus

delay per metre of distance: 3. 33 nanoseconds

$$(1/2\pi)\mathrm{d}\phi/\mathrm{d}f = -3.33\mathrm{ns}$$

$\phi = -20.9f$ with f in gigahertz (or units of 10^9 cycles/second)

Note that, apart from the time delay $\tau = a/2\pi$, Equation 10. 24 is almost the same as Equation 10. 19, plotted in Fig. 10. 7. This is not to be wondered at, since to integrate $e^{+jx}dx$ between symmetrical limits does not differ significantly from integrating $e^{-jx}dx$ between symmetrical limits. e^{+jx} is a vector rotating in the positive geometric sense whilst e^{-jx} rotates in the negative (clockwise) geometric sense.

10.3.2 To a unit-step excitation

The response of our ideal low-pass filter to a unit step is, as before, the integral of the response to a delta function.

$$R_1(t) = \int_{-\infty}^{t} R_0(x)dx \qquad (10.26)$$

Omitting the infinite time delay $a/2\pi$ and noting that the $-\infty$ lower limit of the integral is required when convolving the unit step with the $(\sin 2\pi Ft)/2\pi Ft$

$$R_1(t) = 2AF \int_{-\infty}^{t} \frac{\sin 2\pi Fx}{2\pi Fx} dx \qquad (10.27)$$

$$= 2AF \left[\int_{-\infty}^{0} + \int_{0}^{t} \right] \frac{\sin 2\pi Fx}{2\pi Fx} dx$$

$$= \frac{A}{\pi} \left[\int_{-\infty}^{0} \frac{\sin y}{y} dy + \int_{0}^{t} \frac{\sin y}{y} dy \right]$$

The integrand is an even function, so

$$R_1(t) = \frac{A}{\pi} \left[\int_{0}^{\infty} \frac{\sin y}{y} dy + \int_{0}^{t} \frac{\sin y}{y} dy \right]$$

The second integral is known (Reference 15, page 1, Fig. 1) as the sine integral of t or $\mathrm{Si}(t)$. The first integral is $\mathrm{Si}(\infty) = \pi/2$; thus

$$R_1(t) = A\,[1/2 + (1/\pi)\mathrm{Si}(t)] \text{ with } -\infty < t < \infty \qquad (10.28)$$

The method we have used is, in effect, the convolution method, following Equations 8.13 and 8.14 in Section 8.3. Let us now use the Fourier transform method and use Rule (i) and Equation 9.2 of Chapter 9. This equation tells us that the 'spectrum' of an integration in the time world is $1/p$ or $1/j2\pi f$. The unit step is the time integral of the delta function, so if we divide the spectrum of the ideal low-pass filter's response to the delta function by $j2\pi f$, we shall have the spectrum of the filter's response to a unit step. Thus, this spectrum is simply $A/j2\pi f$, since the filter response is 1 between $-F$ and F. In order to find the time response $R_1(t)$ to the unit step we must calculate the Fourier transform of this spectrum.

$$R_1(t) = A \int_{-F}^{F} \frac{e^{j2\pi tf}}{j2\pi f}\,\mathrm{d}f \qquad (10.29)$$

A difficulty now arises, because the denominator becomes zero when $f = 0$, which it must do on its way from $-F$ to F. We expand the integrand as follows

$$R_1(t) = A \int_{-F}^{F} \frac{\cos 2\pi tf}{j2\pi f}\,\mathrm{d}f + jA \int_{-F}^{F} \frac{\sin 2\pi tf}{j2\pi f}\,\mathrm{d}f$$

The second integral is simply $(A/\pi)\mathrm{Si}(2\pi Ft)$, but the first integral, namely

$$\frac{A}{j2\pi} \int_{-2\pi Ft}^{2\pi Ft} \frac{\cos x}{x}\,\mathrm{d}x$$

becomes infinite when $x = 0$; however, outside this tiny region, the integrand being an odd function, the integral is zero; so noting that when $x \ll 1$, $\cos x = 1$, we need only consider

$$\frac{A}{j2\pi} \int_{-\epsilon}^{\epsilon} \frac{\mathrm{d}x}{x} \quad \text{with } \epsilon \ll 1$$

This is a contour integral whose method of treatment is outside the scope of this book. By choosing an appropriate contour in the complex plane that skirts half-way round the 'pole' (zero value of the denominator x) of the integrand and using the theory of residues we have, finally

$$\int_{-\epsilon}^{\epsilon} \frac{\mathrm{d}x}{x} = j\pi$$

whence

$$R_1(t) = [1/2 + (1/\pi)\mathrm{Si}(2\pi Ft)]\,A \text{ with } -\infty < t < \infty$$

In this particular case, the convolution method does not require a knowledge of contour integration, whereas the Fourier-transform method does.

10.4 THE BASIC SAMPLING FUNCTION

Before making use of the sinc function $(\sin x)/x$ depicted in Fig. 10.7, we must give thought to the time function and spectral representations of the basic sampling function. Such a function is an infinite sequence of delta functions. Imagine that we have a time-varying voltage waveform that we wish to sample at regular intervals in order, perhaps, to measure the 'height' of each sample for conversion into a digital code for the formation of digital words to be transmitted as pulse code modulation (p.c.m.). All we have to do is to multiply the waveform by a sequence of delta functions spaced by such intervals of time as to be capable of reproducing all the frequency components in the waveform that are considered to have a significant value. How frequently should the sampling pulses recur? Consider the sequence

$$S(t) = \sum_{n=-\infty}^{\infty} A\delta(t - nT) \tag{10.30}$$

shown in Fig. 10.10. Short-duration pulses of area A recur with a period T. Since S is a periodic function, we must use the Fourier series rather than the Fourier integral, in order to find its spectrum. This is because the Fourier integral assumes that a spectrum is con-

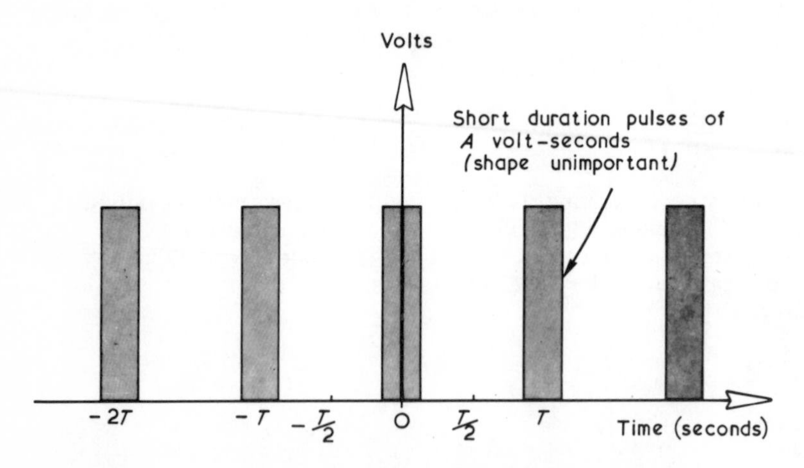

Fig. 10. 10 An infinite sequence of short pulses

tinuous rather than in the form of discrete lines, and this assumption, in turn, arises from the assumption involved in deriving the integral from the series; namely, that a function that is not periodic can be assumed to have an infinite period. Thus, transforming from the series to the integral involves crushing the harmonic intervals denoted by $n = 1, 2, 3$ ..... etc. into infinitesimal steps df of frequency. We use the expressions for the Fourier series given in Chapter 6 and repeated here:

$$\begin{aligned} f(x) = (1/2)b_0 + b_1 \cos x + b_2 \cos 2x + \ldots \\ + b_n \cos nx + \ldots + a_1 \sin x + a_2 \sin 2x + \ldots \\ + a_n \sin nx + \ldots \end{aligned} \tag{10.31}$$

where, to avoid confusion, we substitute u for x inside the integral

$$bn = (1/\pi) \int_{-\pi}^{\pi} f(u)\cos nu.\, du$$

$$a_n = (1/\pi) \int_{-\pi}^{\pi} f(u)\sin nu.\, du \qquad (10.32)$$

with $n = 1, 2, 3$, etc. (i.e. a spectrum of positive frequencies only). This can be simplified if we let n take negative (this is the same thing as making the spectrum display negative frequencies as well as positive ones—a method we have adopted throughout) as well as positive integer values; because whilst $a_0 = 0$ always; b_0 does not, then, have to be shown as being outside the general formula for calculating b_n. First, however, let us combine the oscillating functions of f(x) with their coefficients a_n and b_n.

$$f(x) = (1/2)b_0 + \sum_{n=1}^{\infty} [(1/\pi) \int_{-\pi}^{\pi} f(u)\cos nu \cdot \cos nx \cdot du$$

$$+ (1/\pi) \int_{-\pi}^{\pi} f(u)\sin nu \cdot \sin ux \cdot du] \qquad (10.33)$$

or, better, since the sum of integrals is also the integral of the sum

$$f(x) = (1/2)b_0 + \frac{1}{\pi} \int_{-\pi}^{\pi} f(u) \sum_{n=1}^{\infty} \cos n(u - x)du \qquad (10.34)$$

since $\cos nu.\cos nx + \sin nu.\sin nx = \cos n(u - x)$.

But, from the upper Equation 10.32,

$$(1/2)b_0 = (1/2)\,(1/\pi) \int_{-\pi}^{\pi} f(u)du \qquad (10.35)$$

So, Equation 10.34 becomes

$$f(x) = (1/\pi) \int_{-\pi}^{\pi} f(u)\, [\,1/2 + \sum_{n=1}^{\infty} \cos n(u - x)]\, du$$

Now let n take all integer values between $-\infty$ and $+\infty$, including zero. The cosine of $n(u - x)$ for $n < 0$ will be the same as when $n > 0$, so

$$f(x) = (1/2\pi) \int_{-\pi}^{\pi} f(u) \sum_{n=-\infty}^{\infty} \cos n(u - x)du \qquad (10.36)$$

where the constant term will arise when $n = 0$.

Incidentally, if we note that

$$\cos n(u - x) = (1/2)\,(e^{jnu-x} + e^{-jnu-x})$$

we observe that when $n < 0$, the first term in the parentheses is the same as the second term when $n > 0$ and vice versa, so we may also write Equation 10.36 as

$$f(x) = (1/2\pi) \int_{-\pi}^{\pi} f(u) \sum_{n=-\infty}^{\infty} e^{jn(u-x)} \, du$$

$$= (1/2\pi) \int_{-\pi}^{\pi} f(u) e^{jnu} \, du \sum_{n=-\infty}^{\infty} e^{-jnx} \tag{10.37}$$

The conversion of this expression of the Fourier series to the Fourier integral shown by Equation 6.7 in Chapter 6 is not hard to see. True, the sign of the exponent of the exponential e^{jnu} is positive and that of e^{-jnx} is negative, but they may be reversed at will, since $\cos n(u-x) = \cos n(x-u)$!

Let us return to Fig. 10.10. We are now in a position to calculate the spectrum $S[f]$ of the sequence $S(t)$ of delta functions. Obviously, Equation 10.37 is written in terms of the phase of the oscillating functions; that is, u and x are phase angles measured in radians. Since our function is a time function of period T, the limits of integration in Equation 10.37 become $-T/2$ and $T/2$ and u and x are both to be expressed in terms of the time t as $2\pi t/T$, or $2\pi Ft$ if $F = 1/T$, the fundamental frequency of repetition of the delta functions. Thus Equation 10.37 becomes

$$f(t) = (1/2\pi) \int_{-T/2}^{T/2} f(\tau) e^{jn2\pi\tau/T} d(2\pi\tau/T) \sum_{n=-\infty}^{\infty} e^{-jn2\pi t/T} \tag{10.38}$$

$$= (1/T) \int_{-T/2}^{T/2} f(\tau) e^{jn2\pi\tau T} d\tau \sum_{n=-\infty}^{\infty} e^{-jn2\pi t/T} \tag{10.38a}$$

where τ is a variable of integration to avoid confusion with t in the exponential in the summation sign $\sum$.

Figure 10.10 shows that Equation 10.30 expresses the fact that S is a periodic function whose period is T and for the purposes of Equation 10.38a we shall consider that period which is symmetrical around $t = 0$, because this leads to the simplest result whilst maintaining adequate generality. Thus, substituting $A\delta(\tau)$ for $f(\tau)$, we have

$$S(t) = (1/T)A \int_{-T/2}^{T/2} \delta(\tau) e^{jn2\pi\tau/T} d\tau \sum_{n=-\infty}^{\infty} e^{-jn2\pi t/T} \tag{10.39}$$

Remember that $\int \delta(\tau) d\tau = 1$ when $\tau = 0$ and that it is zero for all other values of $\tau (\neq 0)$. Equation 10.39 becomes

$$S(t) = (A/T) \sum_{n=-\infty}^{\infty} e^{-jn2\pi t/T} \tag{10.40}$$

$$= (A/T)\ [\sum_{n=-\infty}^{\infty} \cos(n2\pi t/T) - j\sum_{n=-\infty}^{\infty} \sin(n2\pi t/T)]$$

and since $\cos(-x) = \cos x$

$\sin(-x) = -\sin x$

we have, finally

$$S(t) = [(A/T)\ 1 + 2\sum_{n=1}^{\infty} \cos(n2\pi t/T)] \qquad (10.40a)$$

$$= (A/T)\sum_{n=-\infty}^{\infty} \cos(n2\pi t/T) \qquad (10.40b)$$

I prefer Equations 10.40 and 10.40b, because they contain implicitly spectra that involve negative as well as positive frequencies and such spectra when transformed back into time functions result in real rather than complex expressions. The engineer usually prefers complex spectra and real time functions rather than the other way round. In the case just treated, however, the transformation between the spectrum with frequencies from $-\infty$ to ∞ to the spectrum with frequencies from 0 to ∞ is quite trivial.

10.4.1 Digression on the Fourier series

Turning now to the interpretation of Equations 10.40, we note, first of all, that they are not spectra, but time functions and this is to be expected, because the Fourier series is a method of representing a time function as a series of sine waves and cosine waves. The spectrum, which is implicit in the series, consists of the coefficients or amplitudes of the expressions of the combined sines and cosines at each harmonic frequency $n2\pi/T$. These coefficients will be functions of the parameters of the periodic waveform and may or may not be functions of the harmonic order n. Suppose a time function $g(t)$ yields the following Fourier series expansion

$$g(t) = \sum_{n=-\infty}^{\infty} [A(k,n)\sin(n2\pi t/T) + B(k,n)\cos(n2\pi t/T)] \qquad (10.41)$$

where A and B are functions of n and of the parameters of $g(t)$, represented by the letter k. We may re-write Equation 10.41 as

$$g(t) = \sum_{n=-\infty}^{\infty} \sqrt{\{A^2(k,n) + B^2(k,n)\}}\ \sin\{[n2\pi t/T + {} + \arctan[B(k,n)/A(k,n)]\} \qquad (10.41a)$$

The spectrum will be three-dimensional and will consist of lines orthogonal to the frequency axis situated at intervals of $2\pi/T$ along it. Each line will have a length equal to $\sqrt{(A2 + B^2)}$ and may be a function of n. The angle of each line, in its own plane orthogonal to the frequency axis, will be arctan (B/A). Such a spectrum will be similar to that shown in Fig. 7.1 in Chapter 7 except that our Fourier-series spectrum is a line spectrum whilst that of Fig. 7.1 is a continuous spectrum. As an example, consider Fig. 10.11 which shows a periodic time function of a rather curious shape. It is simply the addition of a

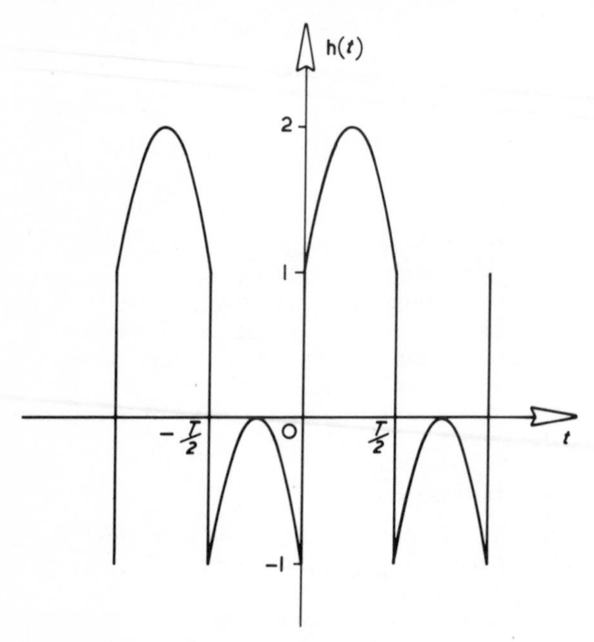

Fig. 10.11 A periodic function of time

simple square wave oscillating between ±1 with period T and a succession of positive-going half-sine waves, (full-wave rectification), also having the period T. The function is, therefore,

$$\left.\begin{array}{l} h(t) = -\sin \omega t - 1 \text{ for } -T/2 < t < 0 \\ \text{and} \\ h(t) = \sin \omega t + 1 \text{ for } 0 < t < T/2 \end{array}\right\} \quad (10.42)$$

where $\omega = 2\pi/T$

The Fourier-series expansion, derived in Appendix 1, produces sine waves and cosine waves, but they never share the same frequency, so we do not have to use Equation 10.41a which would require a more complicated waveform than that shown in Fig. 10.11. The reason that both sines and cosines are required is that h(t) is the sum of an odd and an even function; the half-sine waves in the figure being even, (symmetrical around the ordinate at $t = 0$) and the square wave being odd, (skew-symmetrical around the ordinate at $t = 0$). In spite of the non-sharing of common frequencies by the sines and cosines in the series, it will suffice to indicate the need, in some cases, (this one, in particular) for a three-dimensional spectrum, since the spectral lines that represent the coefficients of sine waves must be drawn at right-angles to the lines that are coefficients of the cosine waves. The Fourier series representing the function in Equation 10.42 is

$$\begin{aligned}h(t) = (2/\pi)\,[\,1 + 2\sin\omega t - (2/3)\cos 2\omega t + (2/3)\sin 3\omega t \\ - (2/15)\cos 4\,\omega t + (2/5)\sin 5\omega t - (2/35)\cos 6\omega t \\ + (2/7)\sin 7\omega t - (2/63)\cos 8\omega t + \ldots\,]\end{aligned} \tag{10.43}$$

Figure 10.12 shows a three-dimensional view of the coefficients of the waves, that is, the spectrum. The d.c. term at $n = 0$ is shown lying in the plane of cosines, because it arises from the cosine series. The $2/\pi$ factor in Equation 10.43 has been omitted from the figure.

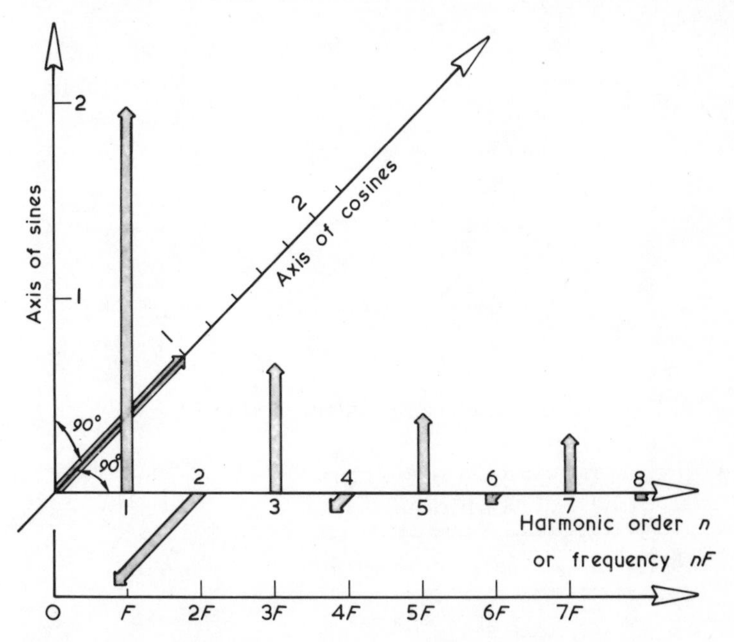

Fig. 10.12 Spectrum of waveform shown in Fig. 10.11

10.4.2 Return to basic sampling function

We are now in a position to discuss, with understanding, the spectrum implicit in the Fourier series expansion of the infinite sequence S(t) of delta functions. Equation 10.40b tells us that the spectrum of S(t), (itself described by Equation 10.30), is a sequence of delta functions in the frequency domain. It is

$$S[f] = (A/T)\sum_{n=-\infty}^{\infty} \delta[f - n/T] \tag{10.40c}$$

Each spectral line, including the one at zero frequency or d.c., has the 'height' A/T. It is a two-dimensional spectrum, because it consists of d.c. preceded and followed by cosine waves only. As far as I know, the only simple periodic time function having a spectrum of the same form as itself is the sequence of delta functions $S(t)$.

10.4.3 Sampling a sine wave

Consider a sine wave of frequency f_1 which we shall sample by means of a sequence of delta functions $S(t)$ of period $T = 1/F$. Let $F > f_1$. All we have to do is to multiply the sine wave $\sin 2\pi f_1 t$ by the sequence (Equation 10.30)

$$S(t) = A \sum_{n=-\infty}^{\infty} \delta(t - n/F)$$

$$= AF \sum_{n=-\infty}^{\infty} \cos n2\pi Ft \qquad (10.40b)$$

whence, denoting the product by $S_1(t)$, we have

$$S_1(t) = AF \sum_{n=-\infty}^{\infty} \sin 2\pi f_1 t \cdot \cos n2\pi Ft \qquad (10.44)$$

$$= AF \sum_{n=-\infty}^{\infty} [(1/2)\sin 2\pi(nF + f_1)t - (1/2)\sin 2\pi(nF - f_1)t] \qquad (10.44a)$$

If we let $n = 0, 1, -1, 2, -2$ etc. we can construct Table 10.2

Table 10.2 SPECTRUM OF INFINITE SEQUENCE OF DELTA FUNCTIONS

n	$(1/2)\sin 2\pi(nF + f_1)t$	$-(1/2)\sin 2\pi(nF - f_1)t$
0	$\frac{1}{2}\sin(2\pi f_1 t)$	$-\frac{1}{2}\sin(-2\pi f_1 t)$
1	$\frac{1}{2}\sin 2\pi(F + f_1)t$	$-\frac{1}{2}\sin 2\pi(F - f_1)t$
-1	$\frac{1}{2}\sin 2\pi(-F + f_1)t$	$-\frac{1}{2}\sin 2\pi(-F - f_1)t$
2	$\frac{1}{2}\sin 2\pi(2F + f_1)t$	$-\frac{1}{2}\sin 2\pi(2F - f_1)t$
-2	$\frac{1}{2}\sin 2\pi(-2F + f_1)t$	$-\frac{1}{2}\sin 2\pi(-2F - f_1)t$

We see that the spectrum consists of lines representing sine waves of value $+1/2$ at the frequencies $\pm nF + f_1$ and $-1/2$ at the frequencies $\pm nF - f_1$. This spectrum is shown between $\pm 3F$ by the solid lines in Fig. 10.13. I have omitted the common factor AF from the figure. Since each line is a delta function of value AF, we can express the spectrum mathematically as

$$S_1[f] = (AF/2) \sum_{n=-\infty}^{\infty} \pm\delta[f - (nF \pm f_1)] \tag{10.45}$$

where the sign in front of the delta function must coincide with the sign before f_1.

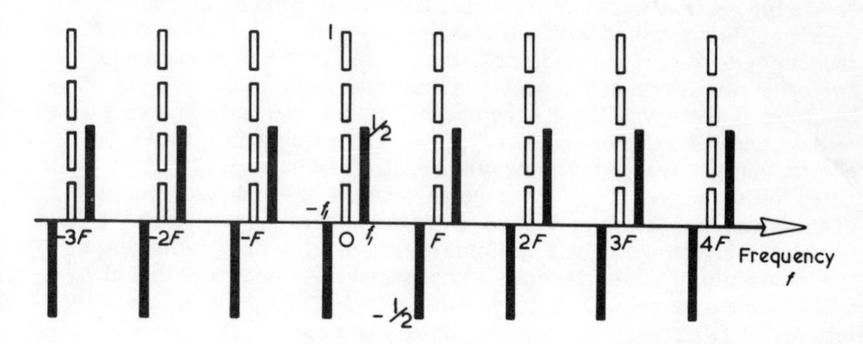

Fig. 10.13 Spectrum of infinite sequence of delta functions

So the process of sampling a sine wave of frequency f_1 leaves the original sine wave undisturbed (spectral lines at $\pm f_1$) on the one hand, but on the other hand it gives rise to replicas of it ($\pm f_1$) situated at harmonics of the sampling frequency (spectral lines at $\pm nF \pm f_1$). These replicas are in fact double-sideband amplitude modulation upon a series of harmonically related suppressed carriers. We can introduce the (suppressed) carriers by adding a d.c. or zero frequency component to our sine wave, $\sin 2\pi f_1 t$. Thus, let our function to be sampled become $1 + m\sin 2\pi f_1 t$ with $m \leqslant 1$. The spectrum of this function is

$$S_2[f] = AF \sum_{n=-\infty}^{\infty} \{\delta[f - nf] \pm (m/2)\, \delta[f - (nF \pm f_1)]\} \tag{10.46}$$

because

$$S_2(t) = S(t) \cdot (1 + m\sin 2\pi f_1 t)$$

and similarly to Equation 10.44:

$$S_2(t) = AF \sum_{n=-\infty}^{\infty} (1 + m\sin 2\pi f_1 t)\cos n2\pi Ft$$

$$= AF \sum_{n=-\infty}^{\infty} (\cos n2\pi Ft + m\sin 2\pi f_1 t \cdot \cos n2\pi Ft)$$

If, to avoid having to re-draw Fig. 10.13, we let $m = 1$, we see that the spectrum, Equation 10.46, is represented by the complete Fig. 10.13, including the solid lines and the dashed lines. The dashed lines are simply the harmonic spectral lines of the sequence of delta functions, but they may also be regarded as the amplitudes of carriers (when $|\pm n| > 0$) having sidebands at $\pm f_1$.

10.4.4. Sampling a periodic waveform; the Nyquist rule

Now suppose that we wish to sample a periodic waveform whose spectrum is contained within a band from $-f_1$ to f_1 including a zero frequency or d.c. component; or if the reader prefers, components of twice the amplitude of the previous ones in a band from 0+ to f_1. The 0+ is to indicate that the d.c. component is not changed. The spectrum of the sampled waveform will now closely resemble Fig. 10.13, except that the frequency intervals $-f_1$ to f_1 and $\pm(nF\pm f_1)$ will be filled with spectral lines corresponding to the frequency components of the spectrum of the original unsampled baseband waveform.

An interesting feature now emerges. Imagine that we increase the bandwidth f_1 whilst leaving F unchanged. The spectral line at f_1 will move to the right, Fig. 10.13, whilst that at F-f_1 will move to the left, and if f_1 exceeds $F/2$ the two lines will cross over each other. If, having sampled our waveform at frequency F, we now wish to recover it by passing the spectrum in Fig. 10.13 through a low-pass filter which will attenuate all the harmonics at $\pm(nF\pm f_1)$, we observe that to avoid passing lower sidebands of the fundamental sampling-frequency term $\pm(F\pm f_1)$ through the filter, we must make the sampling frequency F at least twice the highest frequency f_1 contained in the spectrum of the original waveform. This is Nyquist's sampling theorem. The creation of interference within the passband of the low-pass filter arising from sidebands of the component of fundamental sampling frequency is called 'aliasing'. It can only arise if $f_1 > F/2$.

Before treating the Nyquist sampling theorem using non-periodic time functions, there is an important feature of the sampling of periodic functions which should be considered. Borel's theorem or rule (v) of Chapter 9 tells us that the multiplication of spectra corresponds with the convolution of time functions, Equation 9.8. Although this rule has been given in a chapter dealing with non-periodic functions, I must ask the reader to accept that it applies just as well to periodic functions. Furthermore, due to the fundamental symmetry of the two forms, Equations 6.2 and 6.6, of the Fourier integral (or series, for that matter) we can reverse the statement of rule (v) and say that multiplication of time functions (or any other functions) corresponds with convolution of their spectra. This is particularly easy with periodic functions, since their spectra consist of discrete lines or delta functions. The convolution turns out to be simple arithmetic convolution. Take, for example, the sampling at frequency F of the simple sine wave at frequency $f_1 < F/2$. The spectrum of the repeated delta functions is (see Fig. 10.13) apart from the factor AF:

etc.	$-3F$	$-2F$	$-F$	0	F	$2F$	$3F$ etc.
	1	1	1	1	1	1	1

and the spectrum of the sine wave at f_1 is

$-3F$	$-2F$	$-F$	$-f_1$	0	f_1	F	$2F$	$3F$	
0	0	0	-1	0	1	0	0	0	$\div 2$

So the convolution product of the spectra is, Table 10.3.

10.4.5 A 'burst' of delta functions

Let us now consider the effect upon the spectrum $S[f]$ Equation 10.40c, of restricting the infinite train of delta functions shown in Fig. 10.10, to a finite interval of time. The burst of delta functions will cease to be a periodic function and we may write it as

$$S_k(t) = A\sum_{n=-K}^{K} \delta(t + nT) \tag{10.47}$$

The spectrum $S_k[f]$ will be continuous and will be in terms of density (volts per kilohertz, for example), and no longer will it consist of spectral lines. From Equation 7.5 in Chapter 7

$$\begin{aligned} S_k[f] &= A\sum_{n=-K}^{K} e^{j2n\pi Tf} \\ &= A\sum_{n=-K}^{K} (\cos 2\pi nTf + j\sin 2\pi nTf) \end{aligned} \tag{10.48}$$

But, since $\sin(-x) = -\sin x$, the sine terms for $n < 0$ will be equal in value and opposite in sign from those when $n > 0$; so

$$S_k[f] = A\sum_{n=-K}^{K} \cos 2\pi nTf \tag{10.48a}$$

This is a most awkward spectrum to calculate, because it involves a summation with respect to n at each value of f. As an example we take the following values for the parameters:

$T = 0.1\text{ms}$ $K = 6$

which will make the burst consist of thirteen delta functions, six for $n < 0$, six for $n > 0$ and one in the middle when $n = 0$. Equation 10.48a now becomes

Table 10.3 SPECTRUM OF A SAMPLED SINE WAVE

$-3F-f_1$	$-3F$	$-3F+f_1$	$-2F-f_1$	$-2F$	$-2F+f_1$	$-F-f_1$	$-F$	$-F+f_1$	$-f_1$	0	f_1	$F-f_1$	F	$F+f_1$	$2F-f_1$	$2F$	$2F+f_1$	$3F-f_1$	$3F$	$3F+f_1$	Row No. 1
	1			1			1			1			1			1			1		2
									–1		1										3
–1	0	0	–1	0	0	–1	0	0	–1	0	0	–1	0	0	–1	0	0	–1	0	0	4
0	0	1	0	0	1	0	0	1	0	0	1	0	0	1	0	0	1	0	0	1	5
–1	0	1	–1	0	1	–1	0	1	–1	0	1	–1	0	1	–1	0	1	–1	0	1	6

Row 1: the frequency scale
Row 2: the spectrum of the sampling pulses
Row 3: the spectrum of the function to be sampled. For final answer, divide by 2.
Row 4: the product of row 2 and the left-hand spectral line at $-f_1$ in row 3
Row 5: the product of row 2 and the right-hand spectral line at $+f_1$ in row 3
Row 6: the sum of rows 4 and 5. Each number, (1), should be divided by 2.

$$S_6[f] = A \sum_{n=-6}^{6} \cos(2\pi nf/10) \qquad (10.48b)$$

with f in kHz.

Figure 10.14 shows the spectrum, Equation 10.48b, The spectrum is periodic with a period of 10 kHz.

The definite line-spectrum structure of the spectrum $S[f]$, Equation 10.40c, in Section 10.4.2 no longer exists, but the values of the spectral densities at $f = 0$ and $f = 10$ kHz in Fig. 10.14 certainly predominate and occur, not only at those two frequencies, but at all the multiples of 10 kHz. If we were to increase the number of delta functions in our burst, increasing the time duration of it to suit, we should observe that the spectral densities in Fig. 10.14 at the frequencies in between 0 and 10 kHz (and between all multiples of 10 kHz) would diminish in favour of those close to 0 and 10 kHz and its multiples, and finally, when the burst was of infinite duration, we should be left with a line spectrum.

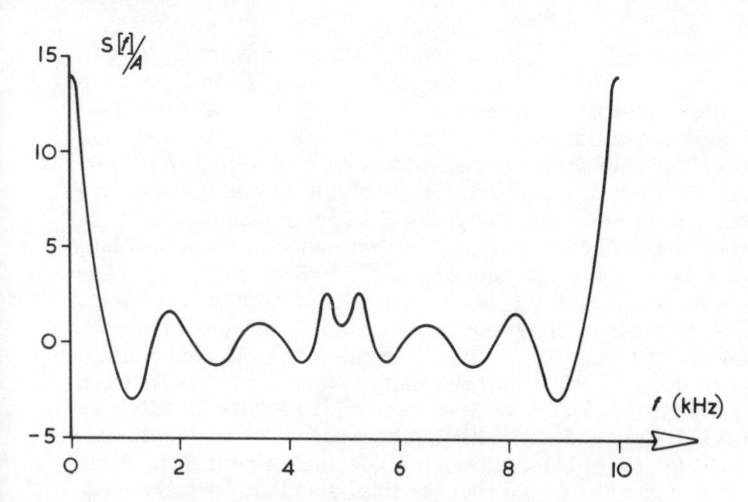

Fig. 10.14 Spectrum of a 'burst' of delta functions

Another point worthy of note is the difference in dimensions between a line spectrum like that resulting from a periodic function, Equation 10.40c, and a continuous spectrum like that given by Equation 10.48a or Fig. 10.14. The magnitude or 'heights' of the lines in a line spectrum have the same dimension as the original function of which they constitute the spectrum. For example, if $S(t)$ is a periodic voltage waveform, its spectral lines $S[f]$ will be measured in volts; but if $S(t)$ is a transient, or non-periodic voltage waveform, the spectrum $S[f]$ will be measured in volts per hertz and is thus a density rather than a discrete quantity. Thus in Equation 10.40c the coefficient A/T has the dimension volts $\times$ seconds/seconds $=$

volts (if the area A is in volts × seconds); whereas in Equation **10.48a** the coefficient A, having the dimension volts × seconds, can be re-written as volts per hertz, which is a voltage density.

10.4.6 Sampling a non-periodic or transient waveform

As an example, let us sample a voltage waveform $v(t)$ which is the response of an ideal low-pass filter to a delta function excitation. If the bandwidth of the low-pass filter is W; that is, we assume it passes without attenuation all frequencies f when $-W > f < W$; then from Equation **10.24** in Section **10.3.1**, we have

$$v(t) = 2AW\,(\sin 2\pi Wt)/2\pi Wt \qquad (10.24a)$$

If now we sample $v(t)$ with a dimensionless version $S'(t)$ of $S(t)$ given by Equation 10.40b and denote the sampled waveform by $v_S(t)$, we have

$$v_S(t) = v(t).\,S'(t) = 2AW\sum_{n=-\infty}^{\infty} \frac{\sin 2\pi Wt}{2\pi Wt}\cos\,(2n\pi t/T) \qquad (10.49)$$

The spectrum $v_S[f]$ of $v_S(t)$ can be obtained immediately by using Borel's theorem or rule (v) of Chapter 9, but reversed in form as we did to construct Table **10.3** in Section **10.4.4**. The product of time functions inside the summation of Equation **10.49** is equivalent to the convolution of their spectra. The spectrum of $2AW\,(\sin 2\pi Wt)\,2\pi Wt)$ is A for frequencies $-W < f < W$ and the spectrum of $\cos\,(n2\pi/T)$ is a sequence of lines of unit height at $f = n/T$. The convolution of these two quantities consists of a rectangular continuous spectrum of height A between $\pm W$ around each frequency $f = n/T$. If the sampling frequency $F = 1/T$ equals or exceeds $2W$, the rectangular spectra will be separate, each from its neighbour. If, on the other hand, $1/T$ is less than $2W$; the rectangular spectra will overlap. For this case the convolution is not quite so obvious. Let us omit the factor A from our spectra and take a case where $F = 1/T = W/3$. Figure 10.15 shows how to effect the convolution. The shaded rectangles represent the spectrum of $\sin 2\pi Wt/2\pi Wt$ for $nF \geq 0$ and the empty rectangles represent the same spectrum for $nF < 0$. To find the final spectrum we add all spectral components within each frequency band 0 to F, F to $2F$ . . . $(n-1)F$ to nF and so on. If we start with the band $-5F$ to $-4F$ we have six, or $2W/F$, sets of components and this holds, up to and including the band $2F$ to $3F$. If I had continued to draw horizontal rectangles centred around $-8F$ to $-\infty F$ and $6F$ to ∞F, it would be plain to see that the final spectrum is rectangular and that its total width depends on how many values of nF we consider. If the delta functions used for sampling were pulses of finite instead of infinitesimal width, the cosine term in Equation **10.49** would change in such a way as to apply progressively increasing attenuation factors to the spectral lines at $\pm nF$ and the vertical width of the rectangles in Fig. **10.15** would decrease as $|\pm n|$ increased, thus reducing the spectral density for increasing frequencies. If the sampled waveform $v_S(t)$ were

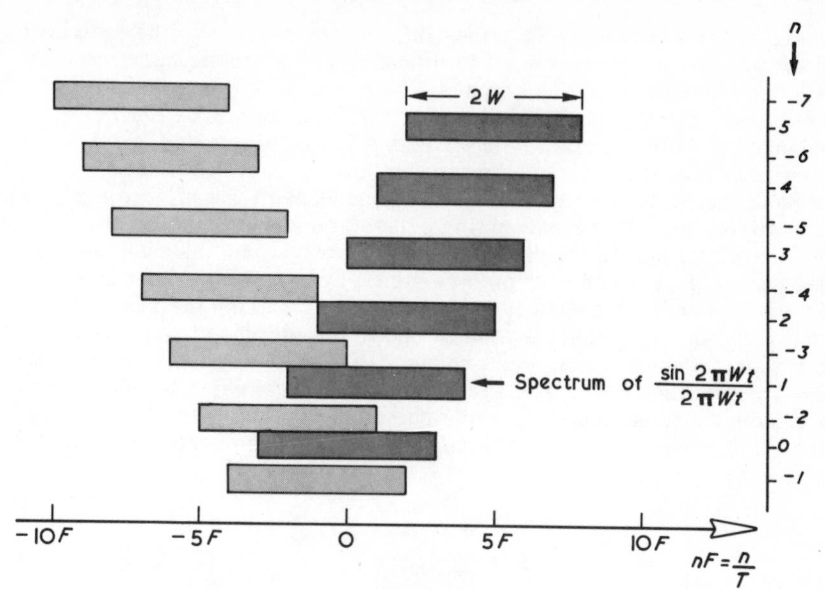

Fig. 10. 15 Convolution involved in finding the spectrum of a sampled sinc-function

inserted into a low-pass filter of cut-off frequency F_1 ($\pm F_1$), the spectrum of the output would be uniform between $\pm F_1$ and zero outside. The example we have considered, wherein $F = w/3$, is in no way a special case and the conclusions we have drawn from it are quite general.

10. 4. 7 Spectrum of a television waveform

The television waveform is an example in which the sampling frequency F, or rather the synchronising-pulse frequency, is much less than the video band W, and Fig. 10. 15 applies. To simplify, we shall consider that the video waveform has neither field (half an interlaced frame or picture) nor frame synchronising pulses, but only line-synchronising pulses. Such a television signal would require a receiver display tube of normal width, but of infinite height. The displayed picture would then resemble a cinematograph film without the black frame-bars that normally separate one frame from its neighbours. For UK 625-line television, the video bandwidth is 5. 5 MHz, so we let $W = 5500$ kHz. Since there are 625 lines per frame or picture and there are 25 frames per second, the line-synchronising-pulse frequency is $F = 625 \times 25 = 15.625$ kHz, and there are exactly 352 spectral lines spaced F apart in the video band of 5500 kHz. The figure is 353 if we include the line at zero frequency.

The line synchronizing pulses that are present in the television waveform do not, in fact, sample that waveform. The process of inserting the synchronising pulses into the waveform may be imagined as follows: first, the waveform is 'blanked' for a certain duration of time once per cycle of F, or what is the same, once per interval of T; the word blanking can be taken to mean 'multiplying by zero'; secondly, the line-synchronising pulses are literally added to the waveform during the already created 'blanking intervals'. It so happens that if we take the direction of increase of the picture signal as positive, then the synchronising pulses are negative-going.

Let us first take a look at the blanking waveform. It must, as already stated, multiply the picture signal by zero for a short interval of time τ once during each line-scan period T. During the remaining part of the line duration we wish to retain the picture signal itself, so the blanking waveform must change its multiplying factor from zero to one, Fig. 10. 16. The synchronising pulses, which we shall add later, are shown as negative-going pulses of width $a < \tau$ and of magnitude -0.4, which is close to the correct value. The spectrum of the

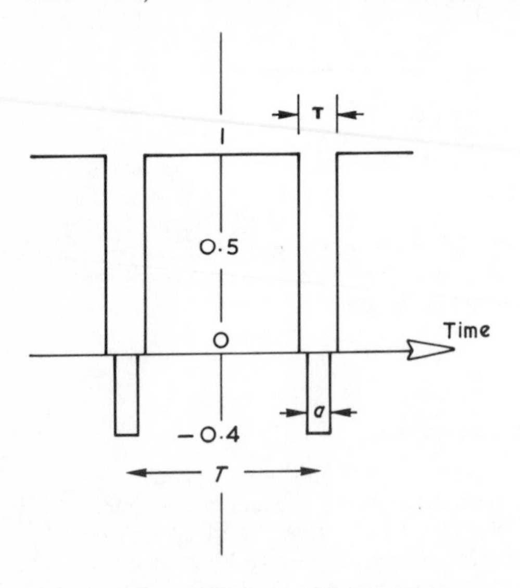

Fig. 10. 16 'Line-blanking' waveform of a television signal

repeated waveform of which one complete period is shown in Fig. 10. 16, may be obtained by using Equation 10. 38a in Section 10. 4. If B(t) is the blanking waveform, assumed to be dimensionless as it will be used simply as a signal modulator,

$$B(t) = 1/T \sum_{n=-\infty}^{\infty} e - jn2\pi t/T \int_{-(T-\tau)/2}^{(T-\tau)/2} 1 \cdot e^{jn2\pi x/T}\, dx$$

Since the limits of integration are situated symmetrically around $x = 0$, the exponential factor becomes a cosine, thus

$$B(t) = 1/T \sum_{n=-\infty}^{\infty} e^{-jn2\pi t/T} (T/2\pi n) \mid \sin (2\pi n x/T) \mid_{-(T-T/2)}^{(T-T)/2}$$

After some manipulation and expanding the exponential factor under the summation sign into cos - j sin, we have finally

$$B(t) = \sum_{n=-\infty}^{\infty} (T-\tau)/T \cdot \frac{\sin n\pi(T-\tau)/T}{n\pi(T-\tau)/T} \cos n2\pi t/T$$

This reveals, once again, a line spectrum, each line having a height given by

$$B[f] = (T-\tau)/T \sum_{n=-\infty}^{\infty} \frac{\sin n\pi(T-\tau)/T}{n\pi(T-\tau)/T} \qquad (10.50)$$

The zero-frequency ($n = 0$) or d.c. spectral line has the value $(T-\tau)/T$; that is, the average value of one period of line-blanking signal. The spectral lines are spaced apart by $F = 1/T$.

The spectrum of the train of synchronising pulses each of width a can be obtained directly from that of the blanking waveform by replacing τ by a and by noting that if the synchronising pulse waveform is denoted by $C(t)$ we may write

$$C(t) = -0.4 \, [1 - B'(t)] \qquad (10.51)$$

in which $B'(t)$ differs from $B(t)$ only in the substitution of a form τ. Inspection of Fig. 10.16 will show the validity of Equation 10.51. Thus

$$C[f] = -0.4 \quad 1 - B'[f]$$

$$= 0.4 \left\{ (T-a)/T \sum_{n=-\infty}^{\infty} \left[\frac{\sin n\pi(T-a)/T}{n\pi(T-a)/T} \right] -1 \right\} \qquad (10.51a)$$

The zero-frequency ($n = 0$) or d.c. term is $-0.4a/T$.

For simplicity of calculation I have shown in Fig. 10.16 the negative-going synchronisation pulses as being placed in the centre of the time duration of the blanking period; that is, a in the middle of τ. In practice, this is not the case; the synchronising pulses are to the left of the position shown in Fig. 10.16 whilst still being wholly inside the time interval τ. For further simplicity, however, I shall now let $a = \tau$, so that the calculation of the spectrum of the complete waveform shown in Fig. 10.16 becomes easy to do. Let us now calculate the spectrum $B[f] + C[f]$ of the waveform shown in Fig. 10.16. If, as proposed above, we let $a = \tau$, we may write

$$D[f] = B[f] + C[f] = -0.4 + 1.4(T-\tau)/T \sum_{n=-\infty}^{\infty} \frac{\sin n\pi(T-\tau)/T}{n\pi(T-\tau)/T} \qquad (10.52)$$

Now, in practice, we have $T = 64$ μs, $\tau = 12$ μs, $a = 4.7$ μs so, since we are going to let $a = \tau$ we shall take the compromise figure of $a = \tau = 8\ \mu s$ whence $\tau/T = 1/8$ and Equation 10.52 becomes

$$D[f] = 1.23 \left[\sum_{n=-\infty}^{\infty} \frac{\sin 2.75n}{2.75n} - 0.326 \right] \tag{10.52a}$$

This is plotted in three separate parts in Fig. 10.17. The ordinate scale is in decibels, $(20\log_{10}D[f])$, because $D[f]$ diminishes from 0.83 to 0.0004 as n ranges from 0 to 352. I have shown values for $n \geq 0$. The spectrum is symmetrical about $n = 0$. The spectral lines occurring at n-equals-multiples-of-eight are zero or $-\infty$ dB. The spacing of the spectral lines is, of course, $F = 1/T$. They are alternately of positive and negative sign and this is indicated by minus (—) signs on the figure. The reader must remember that Fig. 10.17 represents the spectrum of the blanking + line-synchronising pulse waveform of a hypothetical

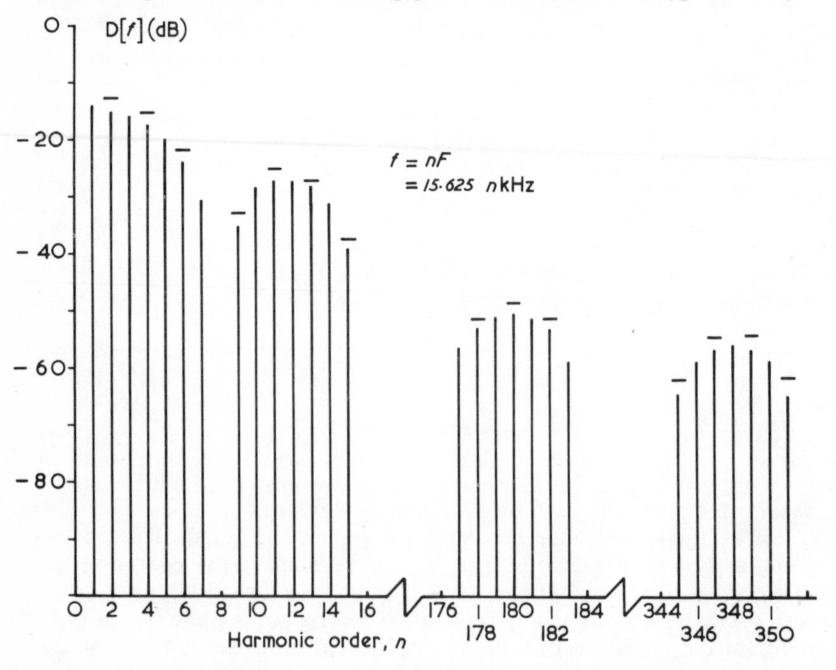

Fig. 10.17 Spectrum of the 'blanking plus synchronising pulses' of a television waveform

signal in which the synchronising pulses are equal in width to the blanking (or zero) portion of the blanking waveform and no field or frame synchronising has been taken into account.

To find the spectrum of a television picture signal submitted to modulation by the blanking waveform to which is later added the synchronising-pulse waveform, we must proceed as follows:

Let the picture signal be denoted v(t). We modulate this with B(t), obtaining v(t)· B(t). We then add the synchronising waveform C(t), obtaining v(t)· B(t) + C(t). If, to simplify, we let $a = \tau$ and therefore B' of Equation 10.51 becomes B, we may write

$$\mathrm{E}(t) = \mathrm{v}(t)\cdot \mathrm{B}(t) - 0.4[1 - \mathrm{B}(t)]$$

$$= -0.4 + [\mathrm{v}(t) + 0.4]\mathrm{B}(t) \qquad (10.53)$$

10.4.7.1 A transient or non-periodic picture signal

Now we shall take v(t) as given by Equation 10.24a in Section 10.4.6, whilst B(t) is given by the equation immediately preceding 10.50, so Equation 10.53 becomes

$$\mathrm{E}(t) = -0.4 + 2AW(1 - \tau/T) \sum_{n=-\infty}^{\infty} \frac{\sin 2\pi WT}{2\pi WT} \cdot \frac{\sin n\pi(1 - \tau/T)}{n\pi(1 - \tau/T)}$$

$$\cos n2\pi Ft + 0.4(1 - \tau/T \sum_{n=-\infty}^{\infty} \frac{\sin n\pi(1 - \tau/T)}{n\pi(1 - \tau/T)} \cos n\, 2\pi Ft \qquad (10.53a)$$

To calculate the spectrum E[f] by the Fourier integral requires very careful interpretation and it is easier to visualise it by the convolution of the spectra of the factors constituting the product of time functions. First, let us analyse Equation 10.53a, The first term −0.4 is a zero-frequency or d.c. term and this must be added to the value of the last indicated summation term when $n = 0$. Secondly, the third and last term represents a spectrum of lines spaced F apart. Thirdly, the second term represents a continuous spectrum obtained by convolution of the type illustrated in Fig. 10.15. Because the spectrum of $(\sin 2\pi Wt)/2\pi Wt$ is a rectangle with vertical ends instead of smoothly attenuating edges, we must, when moving the spectra such as those shown in Fig. 10.15 by each unit change of n, ascribe half the spectrum value (height) to the left-hand value of n and half to the right-hand value. The final spectrum will thus be a smooth function without steps in it. Whenever vertical edges occur, it is reasonable to assume half the value of a function goes with the $x - 0$ part of it and half with the $x + 0$ part of it. E[f] is so complicated if we put real television-system values into it that it is almost impossible to calculate, so we will take a simple example.

Consider, therefore a simple hypothetical television system having the following parameters:

Number of lines per field: $6\frac{1}{2}$
Number of lines per frame: 13
Number of frames per second: 25
Number of picture elements (of duration $1/2W$) per line: 17
(17 is the rounded-off version of $(\frac{4}{3})13$. The factor $\frac{4}{3}$ is the picture aspect ratio: the length of the scan lines is $\frac{4}{3}$ of the height of the picture while this contains as many picture elements per centimetre horizontally as vertically.)

From these figures we obtain $F = 13 \times 25 = 325$ Hz; $T = 0.00377$ s; duration of a picture element: $T/17 = 0.00018$ s and therefore $W = 2760$ Hz; however, so as to ensure a round number (just to make calculation easier, that's all) for the ratio W/F we shall revise the video band to $W = 2600$ Hz hence obtaining $W/F = 8$. We retain the value of the ratio $\tau/T = 1/8$ as before Equation 10.52a. Equation 10.53a becomes

$$E(t) = -0.4 + 4550A \sum_{n=-\infty}^{\infty} \frac{\sin 2\pi Wt}{2\pi Wt} \cdot \frac{\sin 7n\pi/8}{7n\pi/8} \cos 2\pi nFt$$

$$+ 0.35 \sum_{n=-\infty}^{\infty} \frac{\sin 7n\pi/8}{7n\pi/8} \cos 2\pi nFt \qquad (10.53b)$$

where $2W(1 - \tau/T) = 4550$ Hz

The dimension of $E(t)$ is volts, because the -0.4 is the synchronising-pulse magnitude in volts and if A is in volts $\times$ seconds and W in hertz (cycles per second), then the factor $4550A$, which contains W in it, is also in volts. The factor 0.35 being the product of synchronising-pulse magnitude and $(1 - \tau/T)$ is also in volts.

Let us deal, first, with the second and most complicated term in Equation 10.53b. Since $W = 8F$, we must consider rectangular spectra of magnitude $(1 - \tau/T)A$ that are $\pm W = \pm 8F$ wide and we must consider these spectra as sitting on carriers situated, in turn at $0F$, $-1F$, $1F$, $-2F$, $2F$, etc. Since we know from inspection of Equation 10.35b that the final spectrum is symmetrical, we may omit all negative-frequency carriers (and therefore the spectra sitting on them) having frequencies less than $-8F$, because their upper sidebands will be situated in the negative-frequency half of the spectrum space. Figure 10.18 shows the successive spectra of $2AW(1 - \tau/T) \sin 2\pi Wt/2\pi Wt$ shifted one n-unit at a time both horizontally and vertically down the figure. (With $A(1 - \tau/T)$ assumed to have unit value for the time being; remember that the spectrum of $2AW(1 - \tau/T) \sin 2\pi Wt/2\pi Wt$ is simply $A(1 - \tau/T)$ between $\pm W$.) All spectra for which $n < -8$ have been omitted. In fig. 10.18, I have distinguished the spectra having carrier frequencies for which $n \geq 0$ from those with carrier frequencies for which $-8 \leq n < 0$ by writing into each spectrum for $n \geq 0$ its value or magnitude in arbitrary units of volts per hertz. These numbers are really the values of the amplitudes $[\sin(7n\pi/8)]/(7n\pi/8)$, Equation 10.53b,, of the carriers $\cos 2\pi nFt$. If we now add the spectral contributions in each column of n, we have the numbers shown in Fig. 10.18 as 'sums of columns'. These number are plotted as a graph in Fig. 10.19. This is, in relative figures, the spectrum of the blanked television signal consisting of a single, unidimensional, infinitely small, infinitely bright spot in the middle of one scanning line, but assumed to have been band limited by a hypothetical optical system (camera lens) taken as a low-pass filter whose cut-off frequency in cycles per millimetre corresponds to the electronic circuit width $W = 2.6$ kHz. (Hypothetical in two senses: first, that its bandwidth is rectangular and secondly

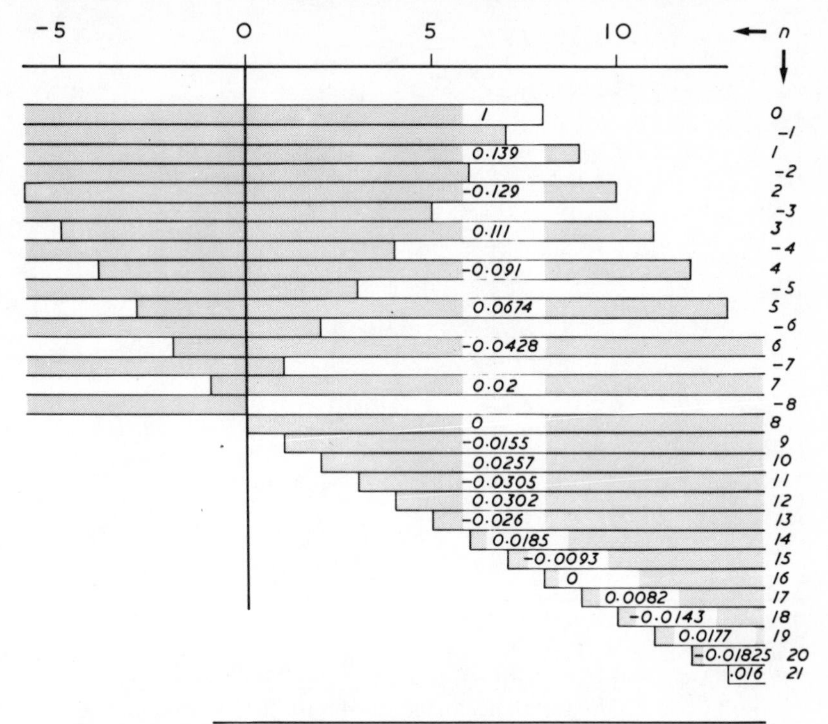

Fig. 10. 18 Convolution involved in finding the spectrum of the television signal of a single white dot

that from a beam of incoherent light (optical random noise if you like) it can produce a (sin x)/x or sinc-function response. We know from Section 10. 2. 2 that this is impossible and that the response would be $|(\sin x)/x|$; that is, all lobes positive.) Note how rapidly the spectrum falls off beyond $n = 8$ or $nF = 8F = W = 2.6$ kHz. The reason that this spectrum does not actually fall to zero above W is because the blanking waveform has vertical edges which demand, for their construction, frequencies ad infinitum.

Before dealing with the addition of this spectrum to those corresponding to the first and third terms of Equation 10. 53b, we must put a value to the quantity $A(1 - \tau/T)$. Well, we know that $1 - \tau/T = 7/8$, but what about A? The blanking waveform in Fig. 10. 16 shows that the picture signal may occupy the amplitude range from 0 to 1, whilst the synchronising pulses occupy the range 0 to -0.4. We may, therefore, so evaluate A that the maximum value of $2AW (\sin 2\pi Wt)/2\pi Wt = 1$. (I am neglecting the negative lobes of the sin x/x function which would invade the synchronising region.) This occurs, of course, for $\tau = 0$,

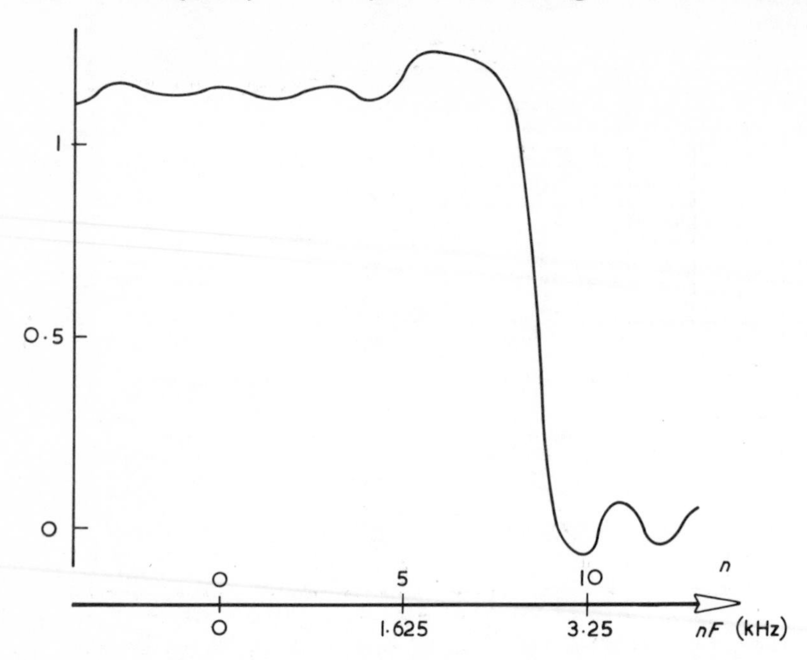

Fig. 10. 19 Spectrum of the modulation product of the blanking waveform and that caused by the picture signal due to a single white dot

whence $A = 1/2W$. So the spectrum in Fig. 10. 19 must be multiplied by $A(1 - \tau/T) = (1 - \tau/T)2W = 7/16W$ or 1.675×10^{-4}. We may now add it to the two other spectra, if we can! In fact we can't, because this spectrum is continuous and the ordinate scale of a graphical representation of it would be in units of volts per hertz, whilst that of the other two terms would be in volts. We may, however, transform the spectrum of Fig. 10. 19 into a histogram by taking the areas under the curve for blocks of frequencies, say 325 Hz wide. The process is then simply to multiply the average ordinate value within a 325 Hz block of spectral densities in Fig. 10. 19 by the product of 1.675×10^{-4}. The amplitudes of the third term in Equation 10. 53b are the same as the amplitudes of the carriers shown inside each rectangular spectrum in Fig. 10. 18 except that we must multiply each one of them by the coefficient 0. 35 shown in Equation 10. 53b. Finally, we have Fig. 10. 20, which shows the spectrum of our hypothetical simplified television system. Although the spectrum is, as already mentioned, a combination of continuous spectral densities and discrete spectral lines we have, by using the histogram principle, enabled the ordinate scale to apply to both forms. If we had included field and frame synchronising pulses, the spectrum would have sprouted lines spaced apart by field (50 Hz) frequency and by frame (25 Hz) frequency. They

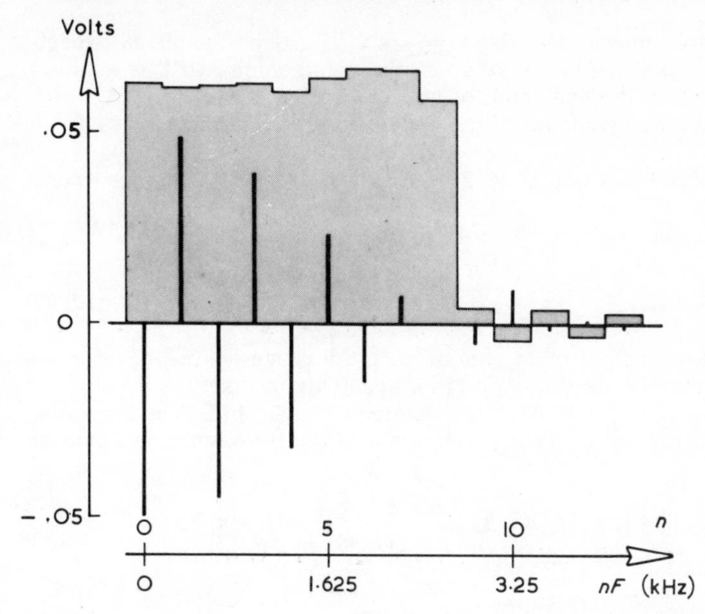

Fig. 10. 20 Spectrum of the complete television signal that would portray a single white dot

would have had much smaller amplitudes, because, occurring less frequently, they would represent much less energy.

10. 4. 7. 2 A repeated picture signal

As an example of a periodic picture signal we might have chosen a simple sine wave, but this would fail to reveal an important point, which is that the spectrum of a repeated function resembles that of a transient or single function identical to the repeated function; the difference lies in the fact that the spectrum of the repeated function is composed of discrete lines instead of being continuous with ordinate units measured in density, (i.e. 'force' per unit of frequency).

We shall, therefore, let $v(t)$, the picture signal, take the form of an infinite succession of sinc functions: $2AW(\sin 2\pi Wt)/2\pi Wt$ repeated at intervals T. This converts the single bright dot of television image into a vertical line on the display screen, since it will reappear in the middle of every scanning line.

This time, to find the spectrum of the repeated sinc functions, we shall convolve time functions and then multiply their spectra; that is, the reverse of the procedure we have been using. The action of repeating $v(t)$ can be achieved by convolving it with an infinite sequence of delta functions.

$$v_R(t) = v(t) * \sum_{n=-\infty}^{\infty} \delta(t - nT) \tag{10.54}$$

The sum of delt functions is the same as S(t) in Equation 10.30 except that I have omitted the factor A, since the voltage units will be supplied by v(t). Now the spectrum of v(t) as a single transient is A with $W < f < W$ and the spectrum of the sequence of delta functions is, from Equation 10.40c in Section 10.4.2, $1/T \sum_{n=-\infty}^{\infty} \delta(f - n/T)$. The spectrum of the *convolution* product v(t) will be the *product* $v_R[f]$

$$v_R[f] = A/T \sum_{n=-WT}^{WT} \delta[f - n/T] \tag{10.55}$$

because the frequency f must remain with the range $-W$ to W. Equation 10.55 tells us that $v_R[f]$ is a spectrum of discrete lines restricted to $-W < f < W$, each line having a height A/T. The equation which, in spectral terms, is the equivalent of the time function Equation 10.53b, is

$$E[f] = -0.4 + v_R[f] * (7/8) \sum_{n=\pi/8}^{\infty} \frac{\sin 7n\pi/8}{7n\pi/8} \delta[f - n/T]$$

$$+ 0.35 \sum_{n=-\infty}^{\infty} \frac{\sin 7n\pi/8}{7n\pi/8} \delta[f - n/T] \tag{10.56}$$

We have now returned to convolving spectra, because in the second term of Equation 10.53b the time functions were multiplied. If, as in the previous section, we let $A = 1/2W$ and $T = 0.00377$ s, we have $A/T = F/2W = 1/16$ and so

$$v_R[f] = 1/16 \sum_{n=-WT}^{WT} \delta[f - n/T],$$

whence

$$E[f] = -0.4 + \left[1/16 \sum_{n=-8}^{8} \delta(f - n/T)\right] * \left[(7/8) \sum_{n=-\infty}^{\infty} \frac{\sin 7n\pi/8}{7n\pi/8}\right.$$

$$\left.\delta[f - n/T]\right] + 0.35 \sum_{n=-\infty}^{\infty} \frac{\sin 7n\pi/8}{7n\pi/8} \delta[f - n/T] \tag{10.56a}$$

The convolution in the second term of Equation 10.56a is similar to that shown in Table 10.3 in Section 10.4.4. Figure 10.21 shows three separate line spectra; the second term of Equation 10.56a, the sum of the first and third terms and the sum of all three terms. Note the similarity between the line spectrum A in Fig. 10.21 and the histogram in Fig. 10.20. The difference between the height of the histogram for $n = 8$ and the height of the spectral line for $n = 8$ is due to the fact that I halved the values of the spectral lines at $n = \pm 8$ in the spectrum of the repeated sinc functions compared with the spectral lines for $-7 \leq n \leq 7$. This is the 'spectral line' equivalent of ascribing half the spectrum height to the lower value of n and half to the higher value of n as mentioned in Section 10.4.7.1. The effect, however, is more noticeable with spectral lines than with a histogram in which changes take place in steps, rather than smoothly. Of course, the line spectrum B in Fig. 10.21 is identical with the line spectrum in Fig. 10.20 since they both represent the spectrum of the synchronising pulses.

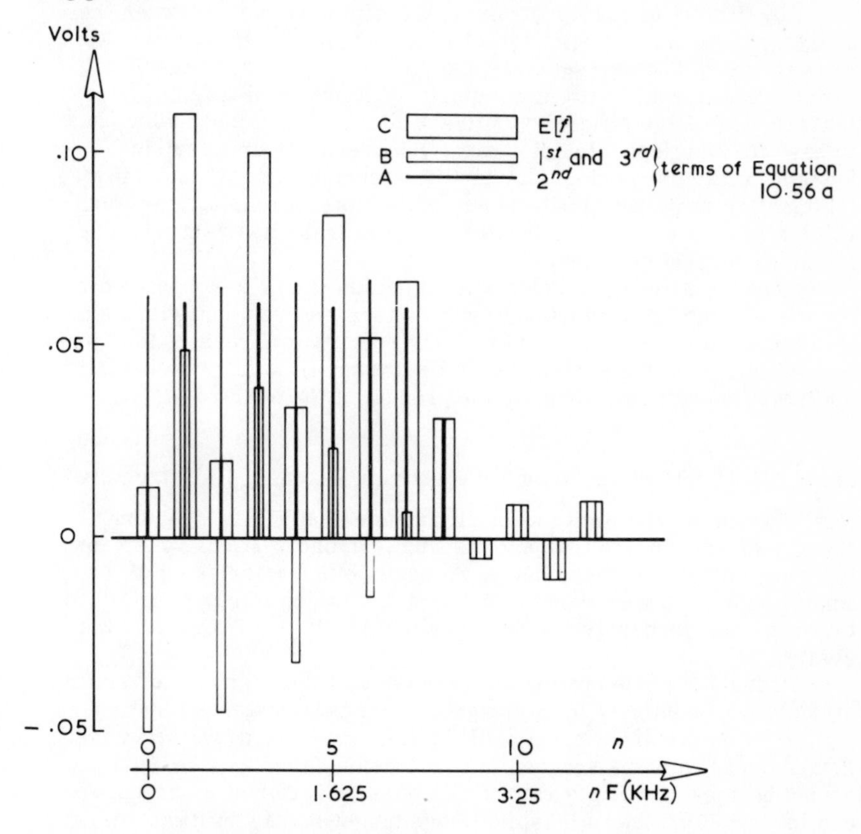

Fig. 10.21 Spectrum of the complete television signal that would portray a single vertical white line

The reader may well ask: 'Why all this concentration upon spectra?'. Well, the principles upon which channels are allocated to various radio services is, up to the present, based on frequency-division multiplex, and this amounts to the juxtaposition of the spectra of the emissions taking place in the various channels so the frequency-planning or channel-allocation engineer wants to know how wide are the spectra of the various signals. In television, it may now be understood by the reader that, for 'still' scenes that do not change with time the spectrum will be composed of discrete lines with unoccupied spaces between them, since the picture signal will be repeated during each frame. If there is movement in the scene each single frame will be a unique transient and so the spectrum will be continuous. In general, apart from 'cuts' from one scene to another, scenic movement is slow and to a first approximation we assume that television spectra are composed of discrete lines. This may be regarded as an inefficient use of spectrum-space, since the gaps between the spectral lines are not used, (or, at least, not much). However, if we were to insert another separate and different television signal in those gaps, we could with minimum, if by no means negligible interference, double the efficiency of use of the frequency channel we had to have for the first television signal. Modern colour television systems make use of this very fact by ensuring that the colouring information required to put colour into a monochrome or black-and-white signal is interleaved between the spectral lines representing the black-and-white signal.

And, of course, for those who do not know how to convolve two (or more) time functions, there is the alternative method, with which we have dealt at length, of finding their spectra (or transfer functions), multiplying them together and then finding the time function having a spectrum of the same form as the product of the spectra.

10.4.7.3 'Vertical spectrum' of television scanning-line structure[16]

For the sake of simplicity we shall assume that all the scanning lines in one television frame are scanned in one vertical sweep down the image display device; that is, we assume that interlacing of the lines in one field with those in the next one is not employed and there is, therefore, only one field per frame (field = frame, for this study).

Consider some optical image or other which, along one vertical straight line, confirms to the brightness-versus-distance function $B(x)$, x, being distance in, say, millimetres measure down the image. If this image is being scanned in a television camera by horizontal scanning lines, say at the rate of 625 lines in 1/25th of a second, we can let $x = 25Ht$ where H is the image height in, say, millimetres. Thus, $B(x) = B(25Ht)$. Now we may, to begin with, imagine the scanning lines to be represented by a succession of delta functions of time viewed as a set of sampling pulses in the vertical direction. I am thinking of the set of points that would appear at the intersections between the scanning lines and a single imaginary vertical line drawn down the optical image. If $1/T (= 15\,625 \text{ Hz})$ is the line-scan frequency, we may write for the sampled signal $B_S(25Ht)$.

$$B_S\,(25Ht) = B\,(25Ht).\sum_{n=0}^{625} \delta(t - nT) \tag{10.57}$$

and if the optical image does not vary with time, all frames will be alike and we can make $-\infty < n < \infty$ instead of $0 \leq n \leq 625$. The spectrum of $B_S(25Ht)$ is, applying Rule (ii), Equation 9.4 of Chapter 9:

$$B_S[f] = (1/25H)B[f/25H]*(1/T)\sum_{n=-\infty}^{\infty} \delta[f - n/T] \tag{10.58}$$

which, as we have seen before, tells us that the 'vertical' spectrum $(1/25H)B[f/25H]$ is repeated at intervals of $F = 1/T$. If this 'vertical' spectrum contains components at frequencies greater than $|\pm F/2|$ there will be interference or aliasing from one scanning line into the adjacent ones due to the overlapping of their spectra. This aliasing will, if the original optical image consists of horizontal black-and-white bands that have a smaller pitch than that of the scanning lines themselves (to take a simple example) cause a further set of horizontal bands to appear in the displayed image. These extra bands will by thicker (vertically) than the scanning lines, because they will have been caused by those frequency components from one double-sideband-carrier system into its neighbours and will therefore be closer to the two neighbouring carriers than the proper limiting frequencies of $\pm F$, which correspond to the scanning-line structure itself. Recall Fig. 10.13 and the discussion that accompanied it. So the first thing to do is to filter the 'vertical' spectrum of the optical image before sampling it with the scanning beam. This can be achieved by optical means, but involves a rather complicated optical system of lenses and spatial filters in the 'Fourier plane' of a lens. It would be easier to shape the 'vertical' profile of the scanning spot into a sinc-function whose passband (in the vertical sense would be rectangular and could be restricted to $\pm F/2 = \pm 1/2T$. Unfortunately, the sinc-function has negative lobes demanding a negative sensitivity in the scanning spot, which being either a beam of light (flying-spot film scanners) or a beam of electrons (cameras) cannot fulfil such a requirement; however, reasonably satisfactory compromises are adopted in practice. The $\cos^2$ function being a favourite. It is so like the Gaussian shape that the difference between them is negligible. So we now know how to stop aliasing from spoiling a television display. Can we eliminate the scanning-line structure? The answer is 'yes', by using the same method as proposed for eliminating aliasing, but this time at the receiving end. We should have to make the 'vertical' profile of the cross-section of the scanning beam in the display tube take the sinc-function form. Again, this is not possible and so display-tube beam profiles have a more or less Gaussian shape which, while tending to blur the achievable vertical resolution, does nevertheless reduce the visibility of the scanning-line structure. As a matter of fact, with European television systems that use 50 fields per second, the lines of each field being interlaced with those of the adjacent or continuous fields, the visibility of the line structure is largely due to the failure of interlace. The lines of one field are reproduced in exactly the same geometric positions on the display tube only 25 times a second and so, unless the viewer sits a long way from the screen, he sees each field of lines separately; that is to say that at 25 frames a second a 625-line interlaced television system is little better than a $312\frac{1}{2}$-line system. The North Americans are better off as they use 60 fields per second, interlaced, and therefore a frame rate of 30 a second. The eye's sensitivity to flicker decreases very rapidly from 25 pictures a second to higher frequencies. In order to eliminate line structure by the method of blurring, the vertical width of each scanning line in an interlaced system (they all are) would have to be $2H/N$ where H is the height of the displayed image and N the number of lines per frame. In 1960, Gibson and Schroeder[17] proposed a better method of eliminating both aliasing and the line structure at the receiving end, although interlacing renders the latter

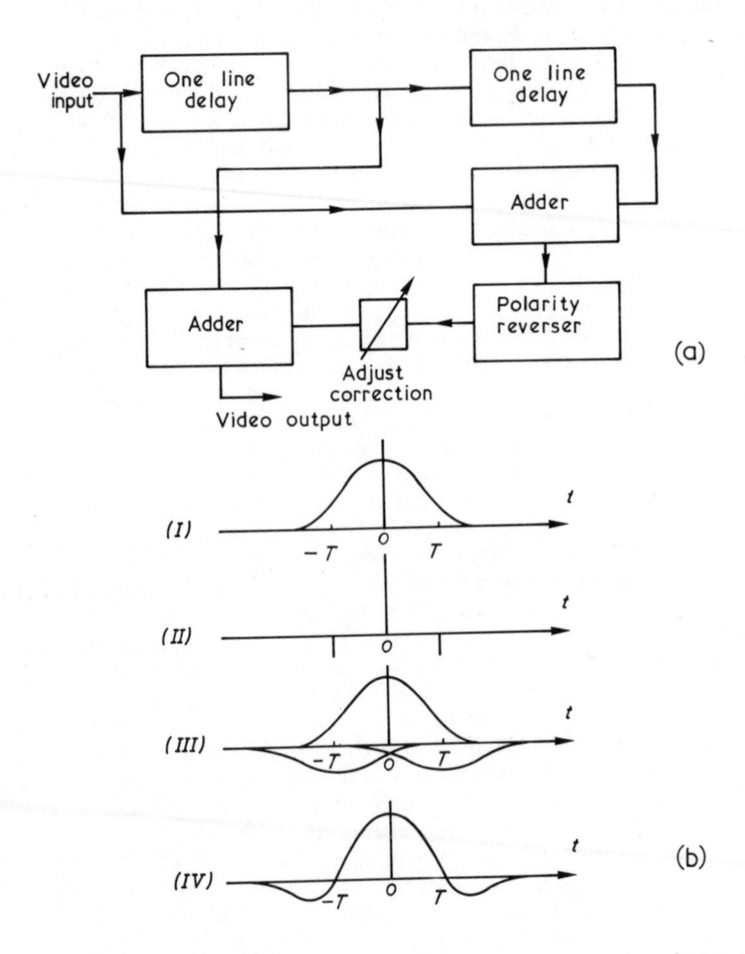

Fig. 10. 22 A television vertical-aperture corrector (BBC figure)

objective less effective than if interlacing were not used. Again, we shall assume no interlacing in order to retain maximum simplicity. Gibson and Schroeder proposed to use line broadening or blurring in the vertical sense at both picture pick-up and at the display and then to effect an aperture-correction system on the electrical signal in between the two ends of the television chain. Figure 10. 22(a) shows the electrical aperture corrector containing two delay lines each having a delay of one scanning-line period T. If we designate the video signal appearing at the junction of the two delay-lines as representing line n of the television signal, then the video input signal at the left of the diagram represents the later line $n + 1$ and the output from the second delay-line represents the line $n - 1$. The device then adds $n - 1$ to $n + 1$, reverses the polarity of the sum and adds a suitable proportion of it to line n. Figure 10. 22(b) shows at (i) a postulated vertical line profile in electrical and time units, at (ii) the three required delta functions which are convolved with the line profile, at (iii) the line profile as convolved and at (iv) the resulting waveform. It can be seen that negative lobes have been achieved and, in theory, if an infinite number of delay lines are used, the complete sinc-function can be simulated and perfection achieved. There remains one snag, however. It is that when the negative lobes are applied directly to the control electrode of the display tube an impossibility is being required from the electron-beam/light-producing-fluorescent-screen complex. This will only occur in the darker regions of the displayed picture, however; since most of the picture detail will occur when the signal has a fairly large direct-current component, so for most of the time the negative lobes will be serving their purpose, because they will be reducing the signal level from one positive value to a lower one during each scanning line. Fortunately, the eye is less sensitive to picture detail in dark patches than in light ones.

The time function corresponding with Equation 10. 57, when a finite vertical aperture is assumed rather than a delta function, is obtained by replacing the sum of delta functions by the convolution product of the vertical-aperture function, say $G(t)$ and the sum of delta functions, thus

$$B_S(25Ht) = B(25Ht)\cdot G(t) * \sum_{n=-\infty}^{\infty} \delta(t - nT) \qquad (10.59)$$

and the spectrum will be

$$B_S[f] = (1/25H)\, B\,[f/25H] * G[f] \cdot (1/T) \sum_{n=-\infty}^{\infty} \delta\,[f - n/T] \qquad (10.60)$$

11. Convolution Division

In Chapter 2 we saw that convolution of two (or more) functions $f(x_i)$ and $h(x_i)$ expressed in terms of a quantised variable x_1, x_2, x_3 etc. was converted into algebraic multiplication of their generating functions $f_g(u^{xi}) \cdot h_g(u^{xi})$ on condition that, finally, we put $u = 1$, but before doing so we used the powers of u to indicate into which columns the various terms of the product were to be placed.

In Chapter 6 we extended convolution to mathematical functions of a continuous variable x, and by Borel's theorem we showed that the convolution of two (or more) functions could be converted into a product of their characteristic functions. The process of reconverting from the product of characteristic functions to the convolution product of the original functions required a Fourier transformation, whereas the conversion from the product of the generating functions to the convolution product of the original functions in the algebraic case required only the equating of u to unit value. In general, therefore, adopting the Fourier-transform notation of parentheses for original functions and square brackets for generating or characteristic functions, we may write

$$f(x) * h(x) \doteq f[z].h[z] \tag{11.1}$$

[also (5.1) and (6.10)]

where, for algebraic convolution (Equation 2.7)

$$f[z] = \sum_{i=1}^{n} z^{x_i} f(x_i)$$

$$h[z] = \sum_{i=1}^{n} z^{x_i} h(x_i) \tag{2.7}$$

and the symbol $\doteq$ becomes $=$ if we put $z = 1$ at the end of the calculation; and for mathematical convolution (Equation 6.2)

$$f[z] = \int f(x)\, e^{-j2\pi zx}\, dx$$

$$h[z] = \int h(x)\, e^{-j2\pi zx}\, dx$$

and the symbol $\doteq$ means 'Fourier transforms of'.

11.1 EXAMPLES OF ALGEBRAIC CONVOLUTION DIVISION

We now inquire as to the meaning of a quotient of generating functions $f[z]/h[z]$.

Let's take a simple example in which the functions $f(x)$, $f[z]$, $h(x)$ and $h[z]$ are as shown in Table 11.1.

Table 11.1

x	$f(x)$	$h(x)$	$f[z]$	$h[z]$	$f[z]/h[z]$	$f(x)*/h(x)$
3		1		z^3		
4		3		$3z^4$		
5		5		$5z^5$	$-z^5$	−1
6		2		$2z^6$	0	0
7					z^7	1
8	−1		$-z^8$		$2z^8$	2
9	−3		$-3z^9$		$3z^9$	3
10	−4		$-4z^{10}$		$3z^{10}$	3
11	3		$3z^{11}$			
12	14		$14z^{12}$			
13	24		$24z^{13}$			
14	28		$28z^{14}$			
15	21		$21z^{15}$			
16	6		$6z^{16}$			
Column no. 1	2	3	4	5	6	7

The symbol $*/$ in the heading of the last column means 'convolution divide by'. The passage from column 6 to column 7 is traversed simply by letting $z = 1$. The algebraic long division of $f[z]$ by $h[z]$ is set out in Appendix 2 for those who have forgotten this process. Although Appendix 2 shows the dividend, the divisor and therefore the quotient in order of ascending powers of z, the same answer would have been obtained if we had written down the division in order of descending powers of z, and since the powers of z tell us into which column each term of the quotient should be placed, no confusion results. The reader will have guessed that I chose f and h such that an exact quotient with no remainder was obtained. If a remainder occurs, one either leaves it as it appears, or one can continue the division ad infinitum. If the dividend and divisor have been written down in order of ascending powers of z the quotient will be expressed in ascending powers of z and vice versa. The convolution-equivalent of the long division given in Appendix 2 is given in Appendix 3. Careful attention to the columns is now required, because we no longer have the column markers z^i to do the job for us.

The reader will now be asking what are the applications of convolution division. There are many, although, because it is less well known than straightforward convolution (-multiplication) it is rarely used. A very simple case could be taken from the example in Chapter 2 in which we considered the random gain errors of two amplifiers in tandem, Fig. 2.1 and Table 2.1. Suppose that the amplifiers were in a circuit by which broadcasting signals were transmitted from a broadcasting authority to a cable transmission-network authority, one amplifier being within the broadcaster's section of the transmission circuit and the other one inside the network of the cable authority. Suppose that the cable authority complained to the broadcaster that gain stability of broadcast signals was inadequate and Fig. 2.1(e), attached to the complaint, was submitted as evidence. Suppose, for example, that the broadcaster was the owner of the amplifier whose performance was described by Fig. 2.1(c). The broadcaster would convolution-divide Fig. 2.1(e) by Fig. 2.1(c) to obtain Fig. 2.1(d), which would reveal the behaviour of the cable authority's amplifier and thus enable informed negotiation to take place. Using generating functions, we may revert to algebraic long division, although it would be just as easy to use convolution division directly. The generating function representing the dividend, Fig. 2.1(e) is, from the last line in Table 2.1

Fig. 2.1(e): $0.067z^9 \quad 0.134z^{10} \quad 0.201z^{11} \quad 0.201z^{12} \quad 0.201z^{13} \quad 0.134z^{14} \quad 0.067z^{15}$

and the divisor is given by the third line in Table 2.1 as

Fig. 2.1(c): $0.33z^4 \quad 0.33z^5 \quad 0.33z^6$

whence the performance of the cable authority's amplifier would be

Fig. 2.1(d): $[0.067z^9 + 0.134z^{10} + 0.201z^{11} + 0.201z^{12} + 0.201z^{13} + 0.134z^{14} + 0.067z^{15}]/[0.33z^4 + 0.33z^5 + 0.33z^6] = 0.2\,[z^5 + z^6 + z^7 + z^8 + z^9]$

Once again, we are not bothered with a remainder, but we must nevertheless give some consideration to this question. I wrote, earlier, that one can obtain the quotient in either ascending or descending powers of z, the position marker for generating functions.

Now, whilst this is true, it is, nevertheless, unimportant. If there is a remainder rather than an exact quotient, it is not the powers, ascending or descending, of z that matter; it is the values of the coefficients of the z's; for if, in continuing the division of the remainder by the divisor, these coefficients diminish in value as the division continues, then the engineer is in a position to stop the division when the remainder at any given stage in the division process has become neglible. If the operation is carried out by direct convolution division of f(x) by h(x), with due care being taken to put the quotient values in their correct columns, it is plain to see that the following types of quotient represent the physical states given below: (If generating functions are used together with ordinary algebraic long division of a dividend by a divisor, both expressed in terms of powers of z, the same remarks will apply to the coefficients of z, since these will be identical to the numbers taking part in the convolution-division.)

(a) Quotient dying away progressively and monotonically; the dynamic system represented by the quotient will be a damped non-oscillatory one.

(b) Quotient increasing indefinitely; an unstable system.

(c) Quotient is an oscillating suite of numbers; the system is oscillatory.

(d) Quotient is an oscillating suite of diminishing numbers; the system is a damped oscillatory one.

The foregoing list was deduced from an article by Professor Tustin of Birmingham University. [18]

11.2 THE SIMULTANEOUS-EQUATION METHOD

From a theoretical viewpoint, the simplest form of convolution division is the taking of a convolution reciprocal. Thus, if we let f(x) = 1 and h(x), the divisor, be a suite of numbers $h_0\ h_1\ \ldots\ h_n$ we may write

$$\begin{aligned} f(x)*/h(x) &= 1*/(h_0\ h_1\ h_2\ \ldots\ ) \\ &= q_0\ q_1\ q_2\ \ldots\ ) \end{aligned} \quad (11.2)$$

If we now work backwards by convolution-multiplying the quotient q by the divisor or h we shall be able to gain some insight into what is possible and what isn't. To take a simple example, let us assume that the divisor has three terms $h_0\ h_1\ h_2$ and the quotient has four terms $q_0\ q_1\ q_2\ q_3$.

Table 11.2

h_0	h_1	h_2			
q_0	q_1	q_2	q_3		
h_0q_0	h_1q_0	h_2q_0			
	h_0q_1	h_1q_1	h_2q_1		
		h_0q_2	h_1q_2	h_2q_2	
			h_0q_3	h_1q_3	h_2q_3
1	0	0	0	0	0

I have written down the answer to this convolution multiplication as 1 0 0 0 0 0, but we must now see whether we can actually achieve this result. To do this we have the following set of simultaneous linear equations, obtained by equating the sum of the terms in each column of Table 11.2 to the appropriate term of the original dividend 1 0 0 etc.

$$\begin{aligned} h_0q_0 &= 1 \\ h_0q_1 + h_1q_0 &= 0 \\ h_0q_2 + h_1q_1 + h_2q_0 &= 0 \\ h_0q_3 + h_1q_2 + h_2q_1 &= 0 \\ h_1q_3 + h_2q_2 &= 0 \\ h_2q_3 &= 0 \end{aligned} \tag{11.3}$$

There are six equations and only four unknowns $q_0\ q_1\ q_2\ q_3$, so the system is over-specified and Equation 11.2 is not solved with a four-term quotient. Nevertheless, let us take the first four equations from the set 11.3 and find the values of $q_0\ q_1\ q_2\ q_3$ which satisfy them. We find

$$q_0 = 1/h_0 \quad q_1 = -h_1/h_0^2 \quad q_2 = h_1^2/h_0^3 - h_2/h_0^2$$

$$q_3 = 2h_1h_2/h_0^3 - h_1^3/h_0^4$$

These values satisfy the first four Equations 11.3 and if we substitue them into the left-hand sides of the last two Equations 11.3 we can write down the real values of the dividend 1 0 0 etc., which actually results from the convolution multiplication shown in Table 11.2. Thus we have

$$\begin{aligned} &1, 0, 0, 0, [3(h_1/h_0)^2\ (h_2/h_0) - (h_2/h_0)^2 - (h_1/h_0)^4], \\ &[2(h_1/h_0)(h_2/h_0)^2 - (h_1/h_0)^3\ (h_2/h_0)] \end{aligned} \tag{11.4}$$

The two terms in square brackets correspond to the last two columns in Table 11.2.

By a miraculous coincidence either the fifth or the sixth term might also be zero, but not both. So we see that we can never convolution-divide the numeric 'one' by a three-column suite of numbers and obtain a four term quotient which, when convolution-multiplied by the divisor gives the numeric 'one', followed by a suite of zeros. Generalising, if the number of terms in the assumed quotient is n, the result of convolution multiplication of it by the divisor will give a post-hoc dividend in which n of the terms are correct. We can now look at this in a more useful way. First, note that in the suite of terms in Expression 11.4 the column numbers reading from left to right, are zero, one, two, three, four and five. If we remember that the subscripts of the h are column markers, like the powers of u if we had been using generating functuions, we see that all the terms inside the first of the square brackets are column-four terms and those inside the second square brackets are column-five terms, as would be hoped! This is, perhaps, not obvious, but note that h_1/h_0 would be in column $(1 - 0) = 1$, and $(h_1/h_0)^2$ would be in column $(1 + 1) = 2$. h_1/h_0 would be in column $(2 - 0) = 2$; thus $(h_1/h_0)^2\,(h_2/h_0)$ would be in column $(2 + 2) = 4$; and so forth.

If we now let ϵ represent all terms in the suite 11.4 which are error terms, that is, the two sets of terms in square brackets, we can write

$$qh = 1 + \epsilon$$

or

$$1/h = q - \epsilon/h$$

which is merely telling us to continue dividing each remainder by the divisor, ad infinitum. The quotient q will be absolutely convergent if the first term h_0 of the divisor is the largest term, (see suite 11.4), If it is not the largest term it may nevertheless be possible to continue the convolution division until the order or number of the columns reached is high enough to be outside the range of interest of the problem in hand even although the actual numerical values obtained for the terms of the quotient in the higher order columns are by no means negligible.

11.2.1 An example of the simultaneous-equation method

Consider the Gaussian error function, half of which is shown in Fig. 11.1,

$$\mathrm{f}(x) = \{1/\sqrt{(2\pi)}\}\ \exp\,(-x^2/2) \tag{11.5}$$

having unit standard deviation and let us suppose that this function represents the overall statistical distribution of errors of some parameter or other of a cascade of electronic equipment. Now suppose that one of the pieces of apparatus has a rectangular statistical distribution of errors, half of which is represented by the dashed-line rectangle in Fig. 11.1. In order to find the shape of the distribution of errors in the rest of the equipment we must convolution-divide the Gaussian curve f (x) by the rectangle h(x). Of course, both curves are assumed to be symmetrical around $x = 0$.

If we take samples of the two curves at abscissae values of −1.6, −1.2, −0.8, −0.4, 0, 0.4, 0.8, 1.2, 1.6, we have the following suites of ordinates:

$$\left.\begin{array}{ll}\text{for f}(x)\text{:} & 0.0075, 0.022, 0.053, 0.010, 0.195, 0.288, 0.374,\\ & \quad 0.4, 0.374, 0.288, 0.195, 0.110, 0.053, 0.022, 0.0075\\ \text{and} & \\ \text{for h}(x)\text{:} & 0.289, 0.289, 0.289, 0.289, 0.289, 0.289, 0.289,\\ & \quad 0.289, 0.289\end{array}\right\} \tag{11.6}$$

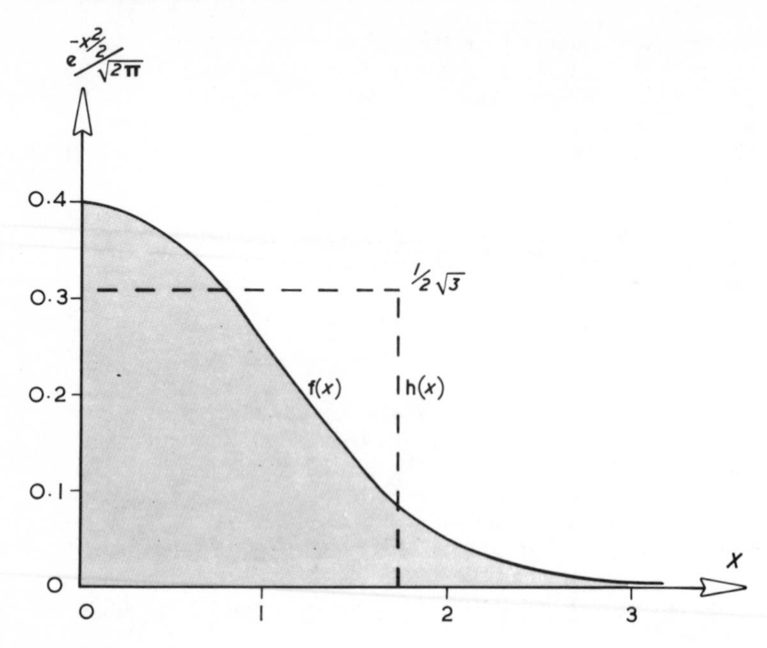

Fig 11.1

Let us assume a quotient of nine terms $q_1 q_2 \ldots q_9$. We now convolve the nine-term unknown quotient with h(x), thus

$$(q_1 q_2 q_3 q_4 q_5 q_6 q_7 q_8 q_9) * 0.289\ (1\,1\,1\,1\,1\,1\,1\,1\,1) = q(x)*h(x)$$

This yields a 17-column convolution-product, Table 11.3. In order to calculate the values of the qs, we are at liberty to choose any nine of the column-sums and if we want a symmetrical quotient q(x)which, when convolved with h(x), the divisor, will reproduce the major portion of the curve f(x), we should choose the middle column of the 17 columns and the four columns each side of it. We then equate the sum of the terms in each of the chosen (double-underlined) columns to the nine chosen values of f(x) and obtain the following nine simultaneous equations:

$$q_1 + q_2 + q_3 + q_4 + q_5 + q_6 + q_7 + q_8 + q_9 = 0.4/0.289 \qquad (11.7a)$$

$$q_1 + q_2 + q_3 + q_4 + q_5 + q_6 + q_7 + q_8 = 0.374/0.289 \qquad (11.7b)$$

$$q_2 + q_3 + q_4 + q_5 + q_6 + q_7 + q_8 + q_9 = 0.374/0.289 \qquad (11.7c)$$

$$q_1 + q_2 + q_3 + q_4 + q_5 + q_6 + q_7 = 0.288/0.289 \qquad (11.7d)$$

$$q_3 + q_4 + q_5 + q_6 + q_7 + q_8 + q_9 = 0.288/0.289 \qquad (11.7e)$$

$$q_1 + q_2 + q_3 + q_4 + q_5 + q_6 = 0.195/0.289 \qquad (11.7f)$$

$$q_4 + q_5 + q_6 + q_7 + q_8 + q_9 = 0.195/0.289 \qquad (11.7g)$$

$$q_1 + q_2 + q_3 + q_4 + q_5 = 0.110/0.289 \quad (11.7h)$$

$$q_5 + q_6 + q_7 + q_8 + q_9 = 0.110/0.289 \quad (11.7i)$$

Because h(x) is rectangular, the above set of equations is quite easy to solve; thus by subtracting Equation 11.7b from 11.7a we have

Equation 11.7a and 11.7b, $q_9 = 0.09$

and similarly, from b and c (and q_9), $q_1 = 0.09$

from c and d (and q_1), $q_8 = 0.29$

from d and e, (and q_1, q_8, q_9), $q_2 = 0.29$

from e and f, (and q_1, q_2, q_8, q_9), $q_7 = 0.325$

from f and g, (and q_1, q_2, q_7, q_8, q_9) $q_3 = 0.325$

from g and h, (and $q_1, q_2, q_3, q_7, q_8, q_9$) $q_6 = 0.295$

from h and i, (and $q_1, q_2, q_3, q_6, q_7, q_8, q_9$) $q_4 = 0.295$

from a (and $q_1, q_2, q_3, q_4, q_6, q_7, q_8, q_9$) $q_5 = -0.62$

Now, referring to Table 11.3, we can fill in line number 14, since we now have numerical values for the qs. Note that line numbers 12 and and 14 are to be multiplied by 0.289 as indicated at the right-hand end of each line. Note, also, that in line 16 I have placed q_5, (= −0.62), in the centre column of f(x), (= 0.4). We see that $q{*}h$, line 15, agrees with f(x), line 13, for the nine values of $-1.6 \leq x \leq 1.6$, but not when x is outside that interval. If we wanted a q(x) such that $q{*}h$ should agree with f(x) from $x = -2.8$ to $x = 2.8$, we should have had to assume 15qs: q_1 to q_{15}. The quotient q(x) of f(x)*/h(x) is shown in Fig. 11.2 in which the values are taken from line 16 of the Table 11.3. It is by no means obvious, at first sight, that the convolution product of a rectangle (dotted lines in Fig. 11.1) with the figure in Fig. 11.2 would yield an approximation to the Gaussian curve shown in Fig. 11.1. If we had effected a straight convolution division of f(x) by h(x) we should have obtained, as quotient

q: 0.0259, 0.05, 0.107, 0.295, 0.322, 0.298, 0.09, −0.09, −0.262, −0.262, −0.187, 0, 0.187, etc., ad infinitum!

which bears remarkably little resemblance to line 16 of Table 11.3 which is at least an approximation to the correct answer. If a convolution division terminates with a non-zero remainder, the 'unknown quotient and simultaneous-equation' method should be used. If the rectangle h(x) had been shrunk to a single ordinate of height 0.289 at $x = 0$, the convolution-division would have become simple arithmetic division of f(x) by 0.289 and a precise quotient would have been obtained. Note, also, that convolution division by a single ordinate of unit height at $x = x_1$, yields a quotient equal to the dividend, but shifted in the direction of diminishing values of x by an amount x_1. This is the natural corollary to the effect of convolution multiplication by unit ordinate at $x = x_1$, as was shown in Fig. 4.2 of Section 4.2.

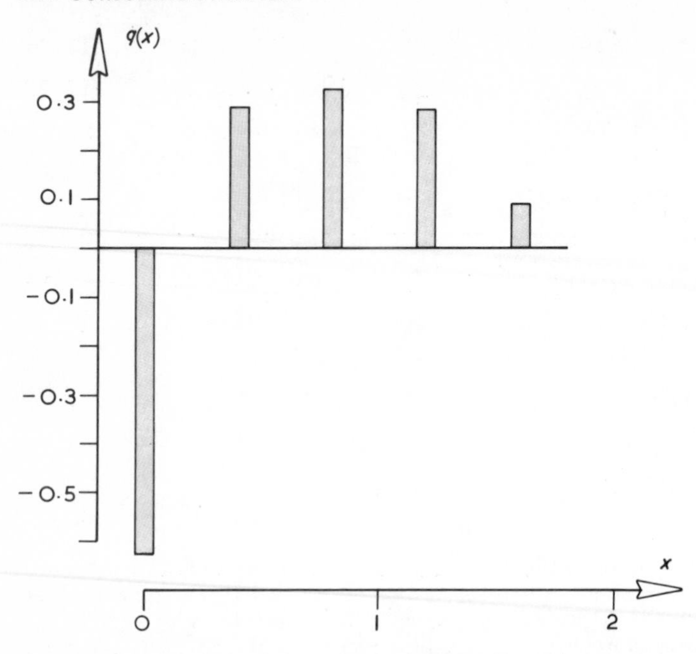

Fig. 11. 2 Approximation to convolution division of solid-line curve by dotted-line rectangle in Fig. 11. 1

11. 3 MATHEMATICAL CONVOLUTION DIVISION OR SERIAL DIVISION

As far as is known there is no explicit mathematical expression equivalent to Equation 4. 8 in Section 4. 2 (convolution of smooth functions), that would enable convolution division of continuous functions to be effected directly. The student is therefore obliged to pass through the intermediate processes of finding the Fourier transforms or spectra of the dividend and of the divisor, effecting normal mathematical division of the former by the latter, thus finding the spectrum of the quotient and then, by a final Fourier transform, finding the required convolution-quotient. Of course, one may perform a convolution multiplication of the reciprocal of the divisor with the dividend, but this is not often a convenient method.

Convolution division is not such a straightforward operation as convolution multiplication. For example, the convolution division of a Gaussian error function by a rectangular function, as treated algebraically in Section 11. 2. 1. ought, one might think, to lend itself to mathematical methods. Thus, the transform of Equation 11. 5 is, by using the method employed in Section 10. 2. 3, for example,

$$f[z] = \exp(-2\pi^2 z^2) \tag{11.8}$$

Table 11.3 CONVOLUTION DIVISION BY THE SIMULTANEOUS-EQUATION METHOD

Line number																			
1	q(x):	q_1	q_2	q_3	q_4	q_5	q_6	q_7	q_8	q_9									
2	h(x):	1	1	1	1	1	1	1	1	1	× 0.289								
3		q_1	q_2	q_3	q_4	q_5	q_6	q_7	q_8	q_9									
4			q_1	q_2	q_3	q_4	q_5	q_6	q_7	q_8	q_9								
5				q_1	q_2	q_3	q_4	q_5	q_6	q_7	q_8	q_9							
6					q_1	q_2	q_3	q_4	q_5	q_6	q_7	q_8	q_9						
7						q_1	q_2	q_3	q_4	q_5	q_6	q_7	q_8	q_9					
8							q_1	q_2	q_3	q_4	q_5	q_6	q_7	q_8	q_9				
9								q_1	q_2	q_3	q_4	q_5	q_6	q_7	q_8	q_9			
10									q_1	q_2	q_3	q_4	q_5	q_6	q_7	q_8	q_9		
11										q_1	q_2	q_3	q_4	q_5	q_6	q_7	q_8	q_9	
12	$q*h$:	q_1	$q_1 + q_2$	$\sum_1^3 q$	$\sum_1^4 q$	etc.	-	-	-	-	-	-	-	etc.	$\sum_9^6 q$	$\sum_9^7 q$	q_8+q_9	q_9	×0.289
13	f(x)	0	0.0075	0.022	0.053	0.110	0.195	0.288	0.374	0.400	0.374	0.288	0.195	0.110	0.053	0.022	0.0075	0	
14	$q*h$:	0.09	0.38	0.705	1	0.38	0.675	1	1.29	1.38	1.29	1	0.675	0.38	1	0.705	0.38	0.09	×0.289
15	$q*h$:	0.026	0.110	0.203	0.289	0.110	0.195	0.289	0.374	0.400	0.374	0.289	0.195	0.110	0.289	0.203	0.110	0.026	
16	q					0.09	0.29	0.325	0.295	−0.62	0.295	0.325	0.29	0.09					
17	x		−2.8	−2.4	−2	−1.6	−1.2	−0.8	−0.4	0	0.4	0.8	1.2	1.6	2	2.4	2.8		

and the transform of the rectangle $g(x) = 1/2\sqrt{3}$ for $-\sqrt{3} < x < \sqrt{3}$ is, (see Section 10.2.2)

$$g[z] = \frac{\sin 2\pi\sqrt{3}z}{2\pi\sqrt{3}z} \tag{11.9}$$

whence, the spectrum of the required convolution-quotient is

$$f[z]/g[z] = \frac{2\pi\sqrt{3}z \exp(-2\pi^2 z^2)}{\sin 2\pi\sqrt{3}z} \tag{11.10}$$

Now this spectrum has poles, that is, it becomes infinite, each time $2\pi\sqrt{3}z = k\pi$ where $k = -\infty, \ldots -2, -1, 1, 2, \ldots \infty$. It may be that the Fourier transform of Equation 11.10 exists, but I have not been able to find it and I doubt its existence!

I am not at all sure that Equation 11.10 tends to zero as z tends to infinity and that is a necessary condition for the existence of its Fourier transform.

A very simple application of the method of division of spectra followed by a transform is to prove that if

$$f(x) = \frac{1}{\sqrt{(2\pi)}\sqrt{(\delta_1^2 + \delta_2^2)}} \exp\{-x^2/2(\delta_1^2 + \delta_2^2)\}$$

and

$$g(x) = \frac{1}{\delta_2\sqrt{(2\pi)}} \exp(-x^2/2\delta_2^2)$$

then

$$f(x)*/g(x) = \frac{1}{\delta_1\sqrt{(2\pi)}} \exp(-x^2/2\delta_1^2)$$

but I will leave this to the reader as an exercise.

12. Convolution Square-roots

Let us take the convolution square root ($*\sqrt{\ }$) of the following suite, S, of numbers: 1 6 17 30 36 30 17 6 1

Table 12.1 shows how the operation is effected and the steps are as follows:

(a) Question: 'what has a square not exceeding 1, the first number of the suite, S?' Answer: 1 has because $1^2 = 1$. Action: write down 1 as the first number of the convolution square root; that is, put a 1 in line 1 column 6. Then write down the 1 again in line 2 column 5.

(b) Now double the 1 from line 2 column 5 and write it down in line 4 column 4.

(c) Now write down in line 3 column 6 the product of that first term of $*\sqrt{S}$, namely 1, and the 1 written into line 2 column 5: $1 \times 1 = 1$. Subtract the 1 in line 3 column 6 from the first term, (1), of S in line 2 column 6. The result is the zero in line 4 column 6.

(d) Bring down the 6 and the 17 from line 2 columns 7 and 8 to line 4.

(e) Question: 'how many times will the 2 in line 4 column 4 go into the 6 of line 4 column 7?' Answer: 'three times'. Write down 3 in line 4 column 5 and also in line 1 column 7 for the second term of $*\sqrt{S}$.

(f) Multiply the 2 and the 3 of line 4 columns 4 and 5 by the 3 of line 1 column 7 and put the products in line 5 columns 7 and 8, respectively.

(g) Subtract each product, 6 and 9 in line 5 from the 'brought-down' 6 and 17 in line 4, thus obtaining 0 and 8 in line 6 columns 7 and 8. Then bring down the 30 and 36 in line 2 columns 9 and 10 to line 6.

(h) Write into line 6 column 3 the 2 from line 4 column 4 and into line 6 column 4 the doubled 3 (= 6) from line 4 column 5.

(i) Question: 'how many times will the 2 from line 6 column 3 go into the 8 from line 6 column 8?' Answer: 'four times'. Write down 4 in line 1 column 8 and in line 6 column 5.

(j) Multiply the 2 6 4 in line 6 by the 4 from line 1 and write the products in line 7 below the 8 30 36 in line 6.

. . .

and so on.

The foregoing instructions worked perfectly, because the suite of numbers S was chosen as the convolution product $\sqrt{S} * \sqrt{S}$. If S is not a perfect convolution square we find ourselves in similar difficulties to those encountered during convolution division.

12.1 EXAMPLE OF A CONVOLUTION SQUARE-ROOT

Consider the Gaussian error function of Equation 11.5

$$f(x) = \{1/\sqrt{(2\pi)}\} \exp(-x^2/2) \tag{12.1}$$

and let us calculate its convolution square-root. We are to find what function convolved with itself gives $f(x)$.

If $f[z] \doteqdot f(x)$ we have

$$f[z] = \exp(-2\pi^2 z^2) \tag{12.2}$$

from Section 11.2.1.

Now $\sqrt{(f[z]} = \exp(-\pi^2 z^2)$ (12.3)

which we can re-write as

$$\sqrt{(f[z])} = \exp(-2\pi^2 (z/\sqrt{2})^2) \tag{12.3a}$$

The transform of this may be obtained directly from Equations 12.1 and 12.2 and using rule (ii), Equation 9.4 of Chapter 9 with $a = 1/\sqrt{2}$.

Thus

$$*\sqrt{\{f(x)\}} \doteqdot \sqrt{\{f[z]\}}$$

$$= (1/\sqrt{\pi}) \exp(-x^2) \tag{12.4}$$

Let us check this by convolving $*\sqrt{\{f(x)\}}$ with itself!

$$I = [*\sqrt{\{f(x)\}}] * [*\sqrt{\{f(x)\}}] \tag{12.5}$$

$$= (1/\pi) \int_{-\infty}^{\infty} \exp(-\tau^2).\ \exp\{-(x-\tau)^2\} d\tau \tag{12.5a}$$

$$= \{\exp(-x^2)/\pi\} \int_{-\infty}^{\infty} \exp\{-(\tau\sqrt{2} - x/\sqrt{2})^2 + x^2/2\} d\tau \tag{12.5b}$$

Table 12.1 CONVOLUTION SQUARE-ROOT

															Line Number
					1	3	4	3	1					$\sqrt{S}$	1
				1	1	6	17	30	36	30	17	6	1	S	2
					1										3
			2	3	0	6	17								4
						6	9								5
		2	6	4		0	8	30	36						6
							8	24	16						7
	2	6	8	3			0	6	20	30	17				8
								6	18	24	9				9
2	6	8	6	1				0	2	6	8	6	1		10
									2	6	8	6	1		11
									0	0	0	0	0		12
1	2	3	4	5	6	7	8	9	10	11	12	13	14		Column Number

$$= \exp(-x^2/2)/\pi\sqrt{2}) \int_{-\infty}^{\infty} \exp(-w^2)dw \qquad (12.5c)$$

where $w = \tau\sqrt{2} - x/\sqrt{2}$.

Finally,

$$I = \{1/\sqrt{(2\pi)}\} \exp(-x^2/2) \qquad (12.5d)$$

$$= f(x) \qquad \text{(Equation 12.1)}$$

12.2 A FURTHER EXAMPLE: WHAT IS THE CONVOLUTION SQUARE-ROOT OF A RECTANGLE?

Expressed in another way, we may ask 'what is that geometric shape which, when convolved with itself, generates a rectangle?' There is no mathematical solution to this question, or should I say that I haven't been able to find one in real terms although a complex solution may exist.

If one assumes a rectangle to be represented by a suite of ordinates each of unit height and one proceeds with the algorithm describing how to obtain Table 12.1, one arrives at the following suite of numbers for a nine-ordinate rectangle:

$*\sqrt{S}$ = 1 1/2 3/8 5/16 35/128 63/256 231/1024 429/2048 6435/32768

The convolution square of $*\sqrt{S}$, or $[*\sqrt{S}] * [*\sqrt{S}]$ is

S = 1 1 1 1 1 1 1 1 1 0.629 0.462 0.348 0.261 0.191 0.132 0.082 0.386

so the nine-ordinate rectangle is accurately reproduced, but there is a tail of eight (= 9—1) ordinates which ought to be zero. This is an interesting example for the mathematically minded, but I know of no practical case which demands the solution to this problem at the present time.

13. Differentiation and Integration by Convolution

Is there a convolution 'operator' by which it is possible to differentiate or integrate a function $f(x)$ expressed as a suite of numbers $f(x_0)\ f(x_1)\ \ldots\ f(x_n)$? The answer is 'yes'![18] First, let us assume that each ordinate $f(x_i)$ is spaced from its neighbours by a constant abscissa value $\Delta x = x_i - x_{i-1}$.

13.1 CONVOLUTION DIFFERENTIATION

Now the derivative D of the function $f(x)$ at the point half-way between x_{i-1} and x_i is the tangent to the curve of $f(x)$ at that point:

$$D = \lim_{\Delta x \to 0} \frac{f(x_i) - f(x_{i-1})}{\Delta x} \tag{13.1}$$

We now ask ourselves 'what suite of numbers when convolved with $f(x)$ gives rise to a suite of numbers each one of which is a difference $f(x_i) - f(x_{i-1})$ between adjacent samples of $f(x)$?' The answer is 1, –1. Let's try it!

Table 13.1 CONVOLUTION DERIVATIVE

a_0	a_1	a_2	a_3	etc.	
1	–1				
a_0	a_1	a_2	a_3	etc.	
	$-a_0$	$-a_1$	$-a_2$	$-a_3$	etc.
a_0	$a_1 - a_0$	$a_2 - a_1$	$a_3 - a_2$	$-a_3$	etc.

Thus, we may re-write Equation 13.1 as

$$D = \frac{(1 - 1) * f(x)}{\Delta x} \tag{13.1a}$$

So the differential operator when applied to a function of a quantised variable x is

$$d/dx = \Delta/\Delta x = *(1,-1)/\Delta x \qquad (13.2)$$

There are two things to remember, however: first, the derivative D, being proportional to the difference between $f(x + x\Delta)$ and $f(x)$ will, in fact, apply to the abscissa value of $x + \frac{1}{2}\Delta x$ and neither to x nor to $x + \Delta x$. Secondly, both the first and last values of the convolution product $f(x) * (1,-1)$ must be discarded as is apparent from an inspection of Table 13.1.

13.1.1 An example of convolution differentiation

A simple example from electrical engineering is the response of a simple resistance-capacitance circuit to a unit-step voltage excitation. If we take the circuit of Fig. 7.2 in Section 7.1, we have, for the ratio of output voltage to input voltage, Expression 7.6.

$$V_2/V_1 = \frac{1}{1 + jRC\omega}$$

and the response to a unit-step excitation of one volt is, from Equation 8.10 in Section 8.2,

$$\frac{v_2}{v_1}(t) = 1 - e^{-t/RC}$$

Now the current i through the shunt capacitor C is

$$i(t) = C\frac{dv_2}{dt}$$

and since $v_1 = 1$ volt, we may differentiate Equation 8.10 to obtain

$$i(t) = (1/R)e^{-t/RC} \qquad (13.3)$$

If we let $R = 1$ ohm and $RC = 1$ s we have

$$i(t) = e^{-t}, i \text{ in ampères and } t \text{ in seconds} \qquad (13.3a)$$

The functions $(v_2/v_1)(t)$ and $i(t)$ from Equations 8.10 and 13.3a are plotted as solid-line curves in Fig. 13.1

Now let's find $i(t)$ by convolution-differentiation of $(v_2/v_1)(t)$. Let $\Delta t = \frac{1}{4}$ s. Table 13.2 shows the method. The variable t is quantised into quarter-second intervals. v_2/v_1, which is the function we require to differentiate, is read off Fig. 13.1 for the quantised values of t. This is shown in line 3 of the table. Next, we convolve with 1, –1 and obtain line 7. We then divide line 7 by $\Delta t = 1/4$ to reach line 8, which is the required derivative, but at values of t half-way in between those given in line 2, as indicated by lines 9 and 10. The agreement between the figures in line 8 and $i(t)$ shown in Fig. 13.1 is so good that a separate curve cannot be plotted.

Table 13.2 DERIVATIVE OF THE RESPONSE OF AN RC CIRCUIT TO UNIT-STEP VOLTAGE EXCITATION IS THE CURRENT

													Line No.
n	0	1	2	3	4	5	6	7	8	9	10		1
$n\Delta t$	0	1/4	2/4	3/4	4/4	5/4	6/4	7/4	8/4	9/4	10/4		2
$\frac{v_2}{v_1}(n\Delta t)$	0	.221	.394	.528	.633	.714	.776	.825	.865	.895	.918	from Fig. 13.1	3
(1, −1)	1	−1											4
convolution	0	.221	.394	.528	.633	.714	.776	.825	.865	.895	.918	etc.	5
		0	−.221	−.394	−.528	−.633	−.714	−.776	−.825	−.865	−.895	−.918	6
$\frac{v_2}{v_1}(n\Delta t)*(1,-1)$	discarded	.221	.173	.134	.105	.081	.062	.049	.040	.030	.023	discarded	7
$i(t) = \frac{v_2}{v_1}(n\Delta t)/\Delta t$ $*(1,-1)$		.884	.692	.536	.420	.324	.248	.196	.160	.120	.092		8
$(n + 1/2)\Delta t$ for		1/8	3/8	5/8	7/8	9/8	11/8	13/8	15/8	17/8	19/8		9
$n =$		0	1	2	3	4	5	6	7	8	9		10

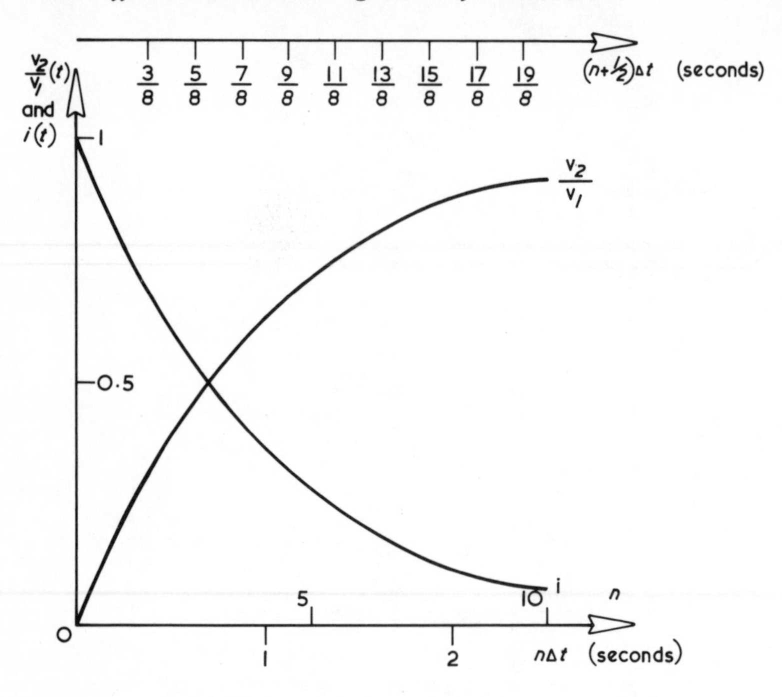

Fig. 13.1 Voltage and current responses of an *RC* circuit to unit-step voltage excitation

The scale of abscissae $n\Delta t$ is repeated at the top of Fig. 13.1 in the form $(n + \frac{1}{2})\Delta t$.

13.2 CONVOLUTION INTEGRATION

This is so simple that it merely requires stating rather than explaining. The process of integration is defined as the limit of the following summation:

$$\sum_{i=-\infty}^{\infty} (x_i - x_{i-1})\mathrm{f}(x_i)$$

in which we let $x_i - x_{i-1} = \Delta x$, when the integral of $\mathrm{f}(x)$ is written

$$\lim_{\Delta x \to 0} \sum \mathrm{f}(x)\Delta x$$

To find the above summation no convolution is required; all we have to do is to multiply by Δx the cumulative total of all the ordinates of f(x) starting at the lower limit of integration and ending with the upper limit. If we want the indefinite integral from a given starting point or given lower limit of integration, we simply calculate the progressive total.

13.2.1 An example of convolution integration

Let us integrate i(t) in Fig. 13.1 and see if we obtain $(v_2/v_1)(t)$. In this particular case there will be no 'constant of integration' to add to the result, because $(v_2/v_1)(0) = 0$. We add each term of i(t) (line 8 in Table 13.2) to the progressive total of all terms to the left, each time multiplying by $\Delta t = 1/4$. Thus, we have Table 13.3, in which the individual ordinates of the integral $\Delta t \sum i(t)$ are, of course, equal to the individual ordinates of $(v_2/v_1)(t)$ as given in line 3 of Table 13.2. Note that the integral of i(t) can have a value only after each time interval Δt has been explored, although the ordinates of i(t) are situated in the middle of each time interval Δt. Thus, whilst i(t) has ordinates $\frac{1}{2}\Delta t$ in advance of those of $(v_2/v_1)(t)$, the integral of i(t), that is $(v_2/v_1)(t)$, has its ordinates back in the original positions, that is at 0, Δt, $2\Delta t$, etc.

13.3 MULTIPLE DIFFERENTIATION AND MULTIPLE INTEGRATION OF FUNCTIONS OF A QUANTISED VARIABLE

There is no need to stress the repeated application of the methods described in Sections 13.1 and 13.2 except to say that they are obviously quite feasible.

Table 13.3 INTEGRAL OF THE CURRENT RESPONSE OF AN RC CIRCUIT TO UNIT-STEP VOLTAGE EXCITATION IS THE VOLTAGE RESPONSE

n		0 to 1	1 to 2	2 to 3	3 to 4	4 to 5	5 to 6	6 to 7	7 to 8	8 to 9	9 to 10
$n\Delta t$ (seconds)		0 to $^1/_4$	$^1/_4$ to $^2/_4$	$^2/_4$ to $^3/_4$	$^3/_4$ to $^4/_4$	$^4/_4$ to $^5/_4$	$^5/_4$ to $^6/_4$	$^6/_4$ to $^7/_4$	$^7/_4$ to $^8/_4$	$^8/_4$ to $^9/_4$	$^9/_4$ to $^{10}/_4$
$i(t)$		.884	.692	.536	.420	.324	.248	.196	.160	.120	.092
	$\Sigma i(t)$	.884	1.576	2.112	2.532	2.856	3.104	3.300	3.460	3.580	3.672
Δt	$\Sigma i(t)$	.221	.394	.528	.633	.714	.776	.825	.865	.895	.918

14. Correlation

It is not an object of this book to make a study of correlation, but as it is a rather similar process to that of convolution I think it advisable to point out the difference between the two operations.

Consider a waveform of random noise arising from thermal agitation of electrons in an electrical circuit, solid-line curve in Fig. 14.1 (by drawing straight lines and joining them with sharp corners I have hopelessly idealized the real situation; for one thing, the sharp corners could only occur if the bandwidth were infinite!); or perhaps better, suppose Fig. 14.1 to be a plot of the daily tempera-

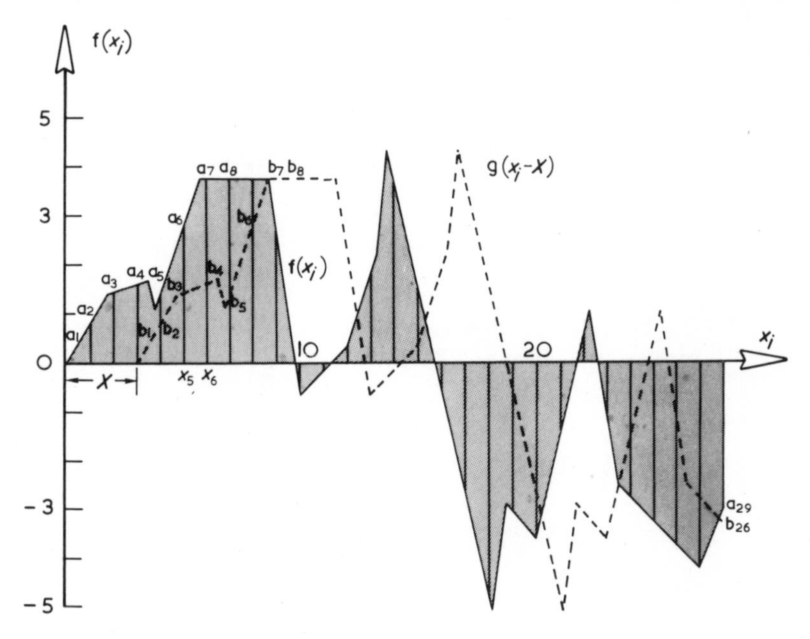

Fig. 14.1 A random function

ture at 24-hour intervals (each ordinate may be considered as the temperature at each midnight, for example) during some leap-year February, the first day being marked as zero on the axis of abscissae. The mean *temperature* defined as the algebraic sum of the ordinates divided by the number of them (29 including the first one of zero height) is 0.184. This is not the mean value of the *abscissae* or first-order moment as defined by Equation 2.10 of Chapter 2; it is the mean of the zero-order moment given by Equation 2.9; that is

$$\frac{1}{n}\sum_{i=1}^{n} f(x_i)$$

Suppose now, we were presented with another 'waveform'—the dashed-line one in Fig. 14.1. How similar are the two graphs? Are they identical? 'Of course they are', you say, but cases that arise are by no means so obvious. The statisticians have proposed a measure of likeness termed 'correlation coefficient'. They suggest that you move progressively along the axis of abscissae and mutliply the ordinate of one curve by the ordinate of the other and then add all the products together and average the result by dividing by the number of products, thus obtaining the correlation coefficient:

$$C = (a_4b_1 + a_5b_2 + \ldots a_{29}b_{26})/26 \quad (14.1)$$

Having done that, what do you know? Almost nothing; but if you move, say, the dashed-line curve, along the axis of abscissae by one abscissa-unit and repeat the operation you have a second correlation coefficient and so on. The function consisting of correlation coefficients with the abscissa values as variable is called the correlation function. The correlation function $C(X)$ for the two graphs or 'waveforms' shown in Fig. 14.1 is plotted in Fig. 14.2 as a function of the amount X of horizontal shift or translation imposed upon the dashed-line waveform. $C(X)$ is maximum when $X = 0$, because it so happens that the two waveforms are really identical and the product of two negative quantities being itself positive, the greatest value for the sum of the products occurs when both graphs are negative at the same abscissae. The algebraic expression for Equation 14.1 is, of course,

$$C(X) = (1/n)\sum_{i=0}^{n} f\{x_i) \cdot g(x_i - X)\} \quad (14.2)$$

Thus, we could re-write Equation 14.1 as

$$C = (a_4b_{4-x} + a_5b_{5-x} + \ldots + a_{29}b_{29-x})/26 \quad (14.1a)$$

if, as was the case in Fig. 14.1, $X = 3$.

Equation 14.2 is deceptively similar to Equation 4.7 in Section 4.2 for the convolution product f*g:

$$P(X) = \sum f(x_i) \cdot g(X - x_i)$$

In fact, the sign of the variable $X - x_i$ in Equation 4.7 is the negative

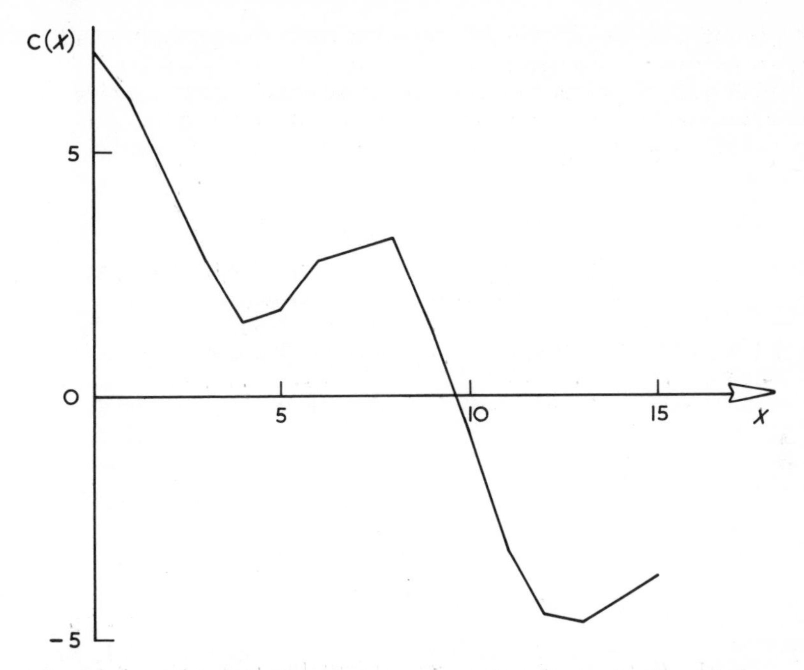

Fig. 14.2 The correlation function of that shown in Fig. 14.1

of that of $x_i - X$ in Equation 14.2; so, reverting to the idea used in Section 4.2 that $X = x + y$, we find that the correlation function could be written

$$C(X) = (1/n) \sum_{i=0}^{n} f(x_i) \cdot g(-y_i) \tag{14.2a}$$

or, using the horizontal bar to indicate 'average',

$$C = \overline{f(x)*g(-y)} \tag{14.2b}$$

So, correlation is 'convolution backwards'! Of course, this makes a considerable difference between the results of the two processes.

14.1 AUTOCORRELATION

Let us now stop pretending, and admit that the two functions shown in Fig. 14.1 are identical, f = g. The correlation function C(X) becomes the autocorrelation function

$$C_a(X) = (1/n) \sum_{i=0}^{n} f(x_i) \cdot f(x_i - X) \tag{14.3}$$

and we see that the autocorrelation function results from multiplying the given function by a delayed version of itself, varying the delay X and adding all the products. It is easy to see that $C_a(X)$ is an even function, for consider the autocorrelation function of the suite of numbers $f(x)$: $-x_2\ -x_1\ x_0\ x_1$. First, let X be a *delay* of one unit, so that

	$-x_2$	$-x_1$	x_0	x_1	0	$f(x)$
	0	$-x_2$	$-x_1$	x_0	x_1	$f(x-X)$
$C_a(X = 1 \text{ unit}) =$	0	x_1x_2	$-x_0x_1$	x_0x_1	0	

and now let X be an *advance* of one unit, so that

	0	$-x_2$	$-x_1$	x_0	x_1	$f(x)$
	$-x_2$	$-x_1$	x_0	x_1	0	$f(x+X)$
$C_a(X = -1 \text{ unit}) =$	0	x_1x_2	$-x_0x_1$	x_0x_1	0	

Thus it is evident that $C_a(X = 1) = C_a(X = -1)$

So, we may now write

$$C_a(X) = 1/n \sum_{i=0}^{n} f(x_i) \cdot f(x_i \mp X) \tag{14.3a}$$

the symbol $\mp$ meaning 'minus or plus', not 'minus and plus'. If we imagine x to be built up from a number of quanta $i\,\Delta x$ we have

$$C_a(X) = 1/n \sum_{i=0}^{n} f(i\Delta x) \cdot f(i\Delta x \mp X) \tag{14.4}$$

Or, if x can be negative as well as positive,

$$C_a(X) = \frac{1}{2n} \sum_{i=-n}^{n} f(i\Delta x) \cdot f(i\Delta x \mp X) \tag{14.4a}$$

The mathematical expression for the autocorrelation function of a continuous function, or one possessing discontinuities of a 'manageable' kind, is, assuming $-N < x < N$

$$C_a(X) = \frac{1}{2N} \int_{-N}^{N} f(x) \cdot f(x - X)\mathrm{d}x \tag{14.5}$$

by the same reasoning as enabled us to pass from Equation 4.7 to Equation 4.8 in Section 4.2. Note that N has the same dimensions as x whereas n was a pure number.

Suppose that the function $f(x)$ has its mean abscissa value $\bar{x}$ at the zero of the x-axis, or if it hasn't, then let $x - \bar{x} = y$ and re-write Equation 14.5 in terms of y

$$C_a(Y) = \frac{1}{2N} \int_{-N-\bar{x}}^{N-\bar{x}} f(y)f(y - Y)dy \tag{14.5a}$$

wherein Y is the analogue of X.

Now let $Y = 0$, whence

$$C_a(0) = \frac{1}{2N} \int_{-N-\bar{x}}^{N-\bar{x}} f^2(y)dy \tag{14.6}$$

This is the mean square value of the function f(y). If the function f(x_i) in Fig. 14.1 represented 28 centimetres-worth of random electrical noise voltage recorded on magnetic tape, then the root-mean-square noise voltage would have been $\sqrt{7}$ volts, (from Fig. 14.2 with $X = 0$).

14.1.1 Spectrum of the autocorrelation function

For simplicity we shall assume that the mean abscissa value of the function f(x) is zero, $\bar{x} = 0$.

Before establishing the spectrum of the autocorrelation function, we must prove two lemmas.

Lemma (i)

The Fourier transform of f($-x$) is f[$-z$]. If, in rule (ii) of Chapter 9 we let $a = -a'$, we should obtain, from Equation 9.4

$$G(-g) = -F[-f]$$

which is incorrect! We must revert to the fundamental transform itself:

$$G(g) = \int_{-\infty}^{\infty} F[f]e^{j2\pi gf} df \tag{14.7}$$

Now let the independent variable of F be $-f$. We have

$$G = \int_{-\infty}^{\infty} F[-f]e^{j2\pi gf} df \tag{14.7a}$$

Let $-f = \lambda$, whence

$$G = -\int_{-\infty}^{\infty} F[\lambda]e^{j2\pi(-g)\lambda} d\lambda \tag{14.7b}$$

Note the reversal of sign in the limits of integration; this did not happen during the development of rule (ii) in the establishment of which the limits of integration became $-a\infty$ and $a\infty$, and, furthermore, there was no change of sign in front of the integral sign. Equation 14.7b may be written

$$G = \int_{-\infty}^{\infty} F[\lambda] e^{j2\pi(-g)\lambda} d\lambda \qquad (14.7c)$$

which is, of course, $G(-g)$!

Lemma (ii)

$f[-z]$ is the complex conjugate of $f[z]$, or if

$$\left.\begin{aligned} &f[z] = a[z] + j\,b[z] \\ &\text{then} \\ &f[-z] = a[z] - j\,b[z] \end{aligned}\right\} \qquad (14.8)$$

Let us, once again, revert to the fundamental Fourier transform

$$F[f] = \int_{-\infty}^{\infty} G(g) e^{-j2\pi fg} dg \qquad (14.9)$$

Now let $G(g)$ be separated into its odd and even parts

$$G(g) = G_0(g) + G_e(g) \qquad (14.10)$$

Figure 14.3 shows three examples of a function G (thick lines), separated into its odd (thin lines) and even (dashed lines) components. To be mathematically correct, by even component is meant a function which has mirror symmetry about the axis of ordinates, whilst odd symmetry means a function having rotational symmetry around the centre (point) of co-ordinates.

We now re-write Equation 14.9 in terms of G_e and G_0 and we also express the complex exponential in terms of its odd and even components, thus

$$F[f] = \int_{-\infty}^{\infty} (G_e + G_0)(\cos 2\pi fg - j \sin 2\pi fg) dg \qquad (14.9a)$$

$$= \int_{-\infty}^{\infty} G_e \cos 2\pi fg \, dg - j \int_{-\infty}^{\infty} G_0 \sin 2\pi fg \, dg +$$

$$+ \int_{-\infty}^{\infty} G_0 \cos 2\pi fg \, dg - j \int_{-\infty}^{\infty} G_e \sin 2\pi fg \, dg \qquad (14.9b)$$

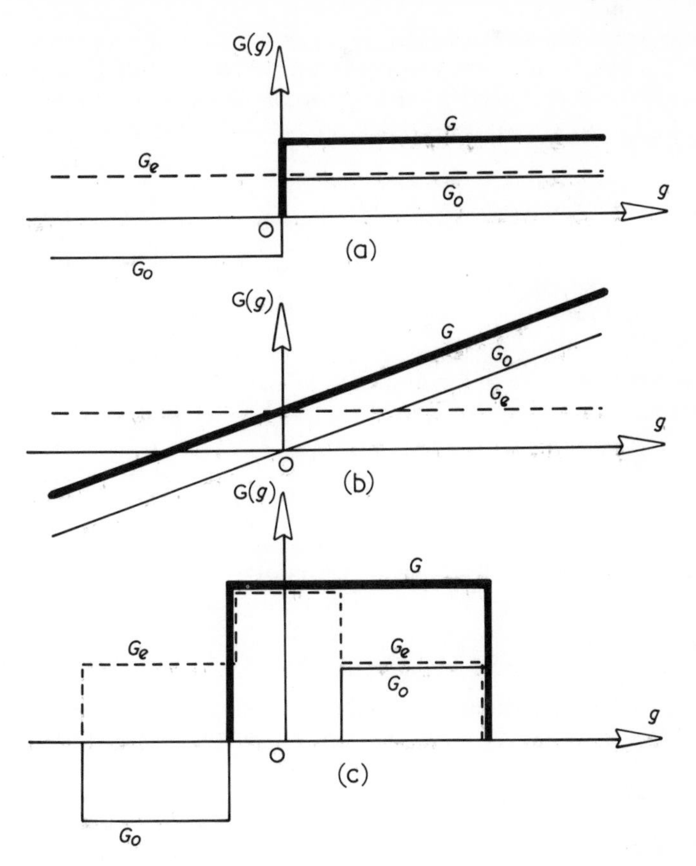

Fig. 14. 3 Odd and even components of each of three functions

The third and fourth integrals are zero, because each integrand being the product of an even and an odd function is itself odd, so that when $g < 0$ the integral becomes the negative of its value for $g > 0$. The first and second integrands may be written

$$F[f] = 2 \int_0^\infty G_e \cos(2\pi fg)dg - 2j \int_0^\infty G_0 \sin(2\pi fg)dg \qquad (14.9c)$$

and, making $f < 0$ we have

$$F[-f] = 2 \int_0^\infty G_e \cos(2\pi fg)dg + 2j \int_0^\infty G_0 \sin(2\pi fg)dg \qquad (14.9d)$$

because $\cos(-a) = \cos a$ and $\sin(-a) = -\sin a$

Thus we see that $F[-f]$ is the complex conjugate of $F[f]$.

Having established the two foregoing lemmas we now return to the search for the spectrum of the autocorrelation function (Equation 14.5)

$$C_a(X) = \frac{1}{2N} \int_{-N}^{N} f(x) \cdot f(x - X)dx$$

First, let $x - X = -x'$, whence $dx = -dx'$ and

$$C_a(X) = \frac{1}{2N} \int_{N+X}^{-N+X} f(-x') \cdot f(X - x')dx' \qquad (14.5b)$$

Since the integral is independent of the variable of integration, we can substitute x for x' in Equation 14.5b and also we can get rid of the minus sign in front by reversing the limits, thus obtaining

$$C_a(X) = \frac{1}{2N} \int_{X-N}^{X+N} f(-x) \cdot f(X - x)dx \qquad (14.5c)$$

This is 'convolution backwards' as mentioned below Equation 14.2b. Thus, in view of lemma (i), the Fourier transform of $C_a(X)$ is the right-hand side of

$$C_a(X) \doteqdot \overline{f[Z] \cdot f[-Z]} \qquad (14.11)$$

where the horizontal bar is used to indicate 'mean value of' instead of retaining the $1/2N$ factor.

$$\text{or} \quad \overline{f(x) \cdot f(x - X)} \doteqdot \overline{f[Z] \cdot f[-Z]}$$

and in view of lemma (ii)

$$C_a(X) \doteqdot | f[Z] |^2 \qquad (14.11a)$$

since

$$\begin{aligned}(a - jb)(a + jb) &= a^2 + b^2 \\ &= | a + jb |^2 \\ &= | a - jb |^2\end{aligned}$$

I have written capital Z in order to bring out the correspondence between X and Z, each having dimensions that are the reciprocal of the other.

It is usual to write Equation 14.5c as

$$C_a(X) = \lim_{N \to \infty} \frac{1}{2N} \int_{N}^{N} f(-x) \cdot f(X - x)dx \qquad (14.5d)$$

which ensures that the integral covers the entire range of values of x required to obtain non-zero products $f(-x) \cdot f(X-x)$. In practice, N is made large enough to satisfy the requirements of the particular problem being investigated. The reader will take note that the factor $1/2N$ converts the integral, which is a sum of products, into the *mean* product.

So finally, the spectrum of the autocorrelation function of a function $f(x)$ is the square of the modulus of its Fourier transform.

14.1.2 The Wiener-Khintchine relation

Relation 14.11a may also be written

$$C_a(X) = \int_{-\infty}^{\infty} f[Z]^2 \; e^{j2\pi XZ} dZ \tag{14.11b}$$

but the modulus of $f[Z]$ is an even function and so is its square, so if we put the exponential term of the integrand into trigonometric form, only the term involving the cosine will contribute to the integral, whence

$$C_a(X) = 2 \int_{0}^{\infty} f[Z]^2 \cos 2\pi XZ \, dZ \tag{14.11c}$$

This is the allegedly well known Wiener-Khintchine relation: 'The autocorrelation function of a time-varying force is equal to the cosine Fourier transform of its power spectrum'. Let us consider, once again, our 28 centimetres-worth of recorded random electrical noise as plotted in Fig. 14.1. Let us, furthermore, assume that its spectrum is $F[f]$* volts/hertz, so that the voltage in a bandwidth df hertz will be $F[f]df$ and in a band 0 Hz to f_0Hz the power will be proportional to

$$V^2 = \int_{0}^{F_0} |F[f]|^2 df \tag{14.12}$$

since the power of a voltage is independent of its phase angle measured against some arbitrary standard, provided that the voltage is assumed to occur across a constant resistance. (I have changed notation from lower case fs:$f(X)$ and $f[Z]$ to capital Fs:$F(t)$ and $F[f]$ to avoid confusion between the function f and the frequency f (in Hertz). Note also that $F[f]$ may well be complex, which would mean that the phase angle of the voltage at frequency f_1 differs from that at f_2, f_1 and f_2 being arbitrarily chosen values.) This assumption is invariably implicit in calculations connected with noise levels and signal-to-noise ratios in communication engineering and broadcasting. We shall return to this question in due course, but suffice it to say, for the present, that to obtain the mean value of $|F[Z]|^2$ we should have to divide it by the time duration of the function $F(x)$ whose autocorrelation function is $Ca(X)$. Thus, in Equation 14.11c X becomes time τ and Z becomes frequency f, and if the spectrum is given in terms of volts/

hertz, whose square is proportional to power, we can see the import of the Wiener-Khintchine relationship:

$$C_a(\tau) = 2 \int_0^\infty \overline{|F[f]|}^2 \cos 2\pi f \tau df \qquad (14.11d)$$

14.1.3 The Wiener-Khintchine relation applied to a sine wave

Let

$$F(t) = A \sin 2\pi f_1 t \qquad (14.13)$$

and let us find the autocorrelation function of $F(t)$. First, we note that $F(t)$ is periodic and a little thought coupled with an inspection of Fig. 14.1 will show that the autocorrelation function of $F(t)$, namely $C_a(\tau)$, will itself be periodic and have the same period as that of $F(t)$. We shall therefore calculate the function $C_a(\tau)$ over two half-periods, one on each side of the axis of ordinates, which is situated at $t = 0$. Thus, using Equation 14.5, we have $N = 1/2f_1$ (because we shall be integrating between $-\pi \leqslant 2\pi f_1 t \leqslant \pi$) and

$$C_a(\tau) = f_1 \int_{-1/2f_1}^{1/2f_1} A \sin 2\pi f_1 t \,.\, A \sin 2\pi f_1 (t - \tau) dt \qquad (14.14)$$

This can easily be shown to be

$$C_a(\tau) = (A^2/2) \cos 2\pi f_1 \tau \qquad (14.14a)$$

which is periodic with period $1/f_1$, like $F(t)$ itself.

The mean-square value of the sine wave, $F(t)$, is $C_a(0) = A^2/2$ leading to the root-mean-square value of $A/\sqrt{2}$, so well known to electrical engineers.

Now, what is the spectrum of $C_a(\tau)$? Well, according to Equation 14.11a, it is the square of the modulus of the Fourier transform $F[f]$ of the original function $F(t)$.

From Chapter 7, Equation 7.4 gives us the Fourier transform of $\cos 2\pi f_1 t + j \sin 2\pi f_1 t$ in the form of a delta function,

$$\exp (j2\pi f_1 t) \doteqdot \delta[f - f_1]$$

Furthermore, it is obvious that

$$\exp (-j2\pi f_1 t) \doteqdot \delta[f + f_1]$$

So by rule (iii), Equation 9.6 in Chapter 9, we may write

$$\delta[f - f_1] - \delta[f + f_1] \doteqdot 2j \sin 2\pi f_1 t \qquad (14.15)$$

and finally, remembering that $F(t) = A \sin 2\pi f_1 t$

$$F[f] = A/2j \,\{\delta[f - f_1] - \delta[f + f_1]\} \qquad (14.16)$$

The spectrum of our sine wave consists, therefore, of two spectral lines one at $f = -f_1$ and the other at $f = f_1$ and each having a height

$-A/2$ and $+A/2$ respectively, lying parallel to the imaginary axis. The squared modulus $|F[f]|^2$ of such a spectrum is $(A^2/4)\delta[f - f_1] + (A^2/4)\delta[f + f_1]$. Note that I have avoided doing the obvious thing, namely squaring the binomial inside the braces $\{\}$ of Equation 14.16. This would have made explicit the cross-product of the two delta functions, which is, of course, zero; because when either one of them is 'active', at f_1 or $-f_1$ the other, being active at $-f_1$ or f_1, will be zero. However, also involved would have been the square of each delta function and that would have involved us in a new kind of delta function whose area would have been infinite in value instead of having unit value. It is evidently not admissible to raise a delta function to a power (such as squaring). This is an interesting point, worthy of note. The delta function is an operator whose principle function is that of a time- or frequency-marker, and whilst it may be differentiated or integrated, it seems to me that multiplication of one delta function by another occurring at the same epoch or the same frequency is inadmissible. In any case, we shall not venture into that field, but will fall back upon physical interpretations rather than attempt to discover a form of mathematics to cope with such situations.

Let us now see whether the cosine Fourier transform of $|F[f]|^2$ is, in fact, equal to $C_a(\tau)$.

We may write

$$\int_0^\infty |F[f]|^2 \cos 2\pi f\tau \, df = A^2/4 \int_0^\infty \delta[f - f_1] \cos 2\pi f\tau \, df$$

$$+ A^2/4 \int_0^\infty \delta[f + f_1] \cos 2\pi f df \qquad (14.17)$$

But, since the two delta functions exist only when $f = \pm f_1$ we have

$$\int_0^\infty |F[f]|^2 \cos 2\pi f\tau df = (A^2/4)\left\{\cos 2\pi f_1\tau \int_0^\infty \delta[f - f_1]df\right.$$

$$\left. + \cos 2\pi(-f_1)\tau \int_0^\infty \delta[f + f_1]df\right\} \qquad (14.17a)$$

$$= (A^2/2 \cos 2\pi f\tau$$

which is $C_a(\tau)$ as given earlier. In this case we didn't have to use the mean-value concept for $|F[f]|^2$, because the function $F(t)$ was periodic—lasting forever and having a line spectrum. Also, we dropped the factor 2, which appears before the integral in Equation 14.11d, because we include in Equations 14.17 and 14.17a the delta function occurring at the negative frequency of $f = -f_1$ as well as the delta function at the positive frequency of $f = f_1$.

14.1.4 The Wiener-Khintchine relation applied to a transient

Consider once again (Section 8.4 and Fig. 7.2 in Section 7.1) the RC circuit and let us calculate the autocorrelation function of its response to a delta function excitation. From Equation 8.16 we have (using our present notation of $\mathrm{F}(t)$ and $\mathrm{F}[f]$)

$$\mathrm{F}(t) = (A/RC)\mathrm{e}^{-t/RC} \quad t \geqslant 0$$

where A is the delta function area or value in, say, volt-seconds and (Equation 7.6)

$$\mathrm{F}[f] = \frac{A}{1 + 2\pi RCf}$$

and

$$|\mathrm{F}[f]|^2 = \frac{A^2}{1 + 4\pi^2 R^2 C^2 f^2}$$

Now, from Equation 14.5, we have, restricting ourselves to $t > 0$,

$$\mathrm{C_a}(\tau) = A^2 \int_0^N (1/RC)\mathrm{e}^{-t/RC} \,.\, (1/RC)\mathrm{e}^{-(t-\tau)/RC}\mathrm{d}t \qquad (14.18)$$

and with $N \rightarrow \infty$

$$\mathrm{C_a}(\tau) = (A^2/2RC)\mathrm{e}^{\tau/RC} \qquad (14.18\mathrm{a})$$

The mean value of the integral in Equation 14.18 taken over the total duration of the function $\mathrm{F}(t)$, namely $0 \leqslant t \leqslant \infty$ would be zero, so we dispense with the idea and simply calculate the area under the product curve. Let us now apply the Wiener-Khintchine theorem. The cosine Fourier transform of the squared modulus is, using Equation 14.11d,

$$2A^2/(RC)^2 \int_0^\infty \frac{\cos 2\pi f\tau \mathrm{d}f}{(1/RC)^2 + 4\pi^2 f^2} \qquad (14.19)$$

or, if we let $x = 2\pi RCf$,

$$2A^2/2\pi RC \int_0^\infty \frac{\cos(\tau x/RC)}{1 + x^2}\mathrm{d}x \qquad (14.19\mathrm{a})$$

This integral (no. 490, p. 63 in Reference 5) is equal to

$$(A^2/RC)\mathrm{e}^{-|\tau/RC|} \qquad (14.19\mathrm{b})$$

which disagrees with Equation 14.18a. Obviously 14.19b is correct, since the exponential exponent ensures (i) that the autocorrelation function is an even one and (ii) that it has a maximum when $\tau = 0$. Equations 14.18 and 14.18a can be rectified if we remember that since Equation 14.5 derived from Equation 14.4a, it could be written

$$C_a(X) = \frac{1}{2N} \int_{-N}^{N} f(x) \cdot f(x \mp X) dx$$

If we take the plus (+) sign in the second parentheses of the integrand we may change the sign of τ in Equations 14.18 and 14.18a, thus achieving

$$C_a(\tau) = (A^2/2RC)e^{-\tau/RC}$$

This is fine so long as $\tau > 0$ and in order to cope with the possibility of having $\tau < 0$ we must use the modulus or absolute-value symbol $|\ |$, whence

$$C_a(\tau) = (A^2/RC)e^{-|\tau/RC|}$$

as in Equation 14.19b. This example constitutes a warning that sometimes the cosine transform of the spectrum modulus is a safer way of calculating an autocorrelation function than the more direct method using Equation 14.5.

The integral of the squared voltage output from the RC circuit; not the mean squared voltage; is, of course,

$$C_a(0) = A^2/RC \text{ in volts}^2 \times \text{seconds}$$

14.1.5 The Wiener-Khintchine relation applied to a rectangular pulse

Let

$$F(t) = A \quad \text{with} \quad 0 < t < T \tag{14.20}$$

The autocorrelation function is

$$C_a(\tau) = 1/T \int_0^T A(t)\, A(t - \tau) dt$$

Figure 14.4 shows immediately that (14.21)

$$C_a(\tau) = (1/T)A^2\,(T - |\tau|)$$

$$= A^2\,(1 - |\tau/T|) \tag{14.22}$$

It is necessary, either to restrict τ to positive values only, or to use the 'absolute value' sign $|\ |$ to ensure that when $\tau < 0$ the autocorrelation function, being an even function, has the same values as when $\tau > 0$.

The spectrum of F(t) is

$$F[f] = A \int_0^T e^{-j2\pi ft}\, dt \tag{14.23}$$

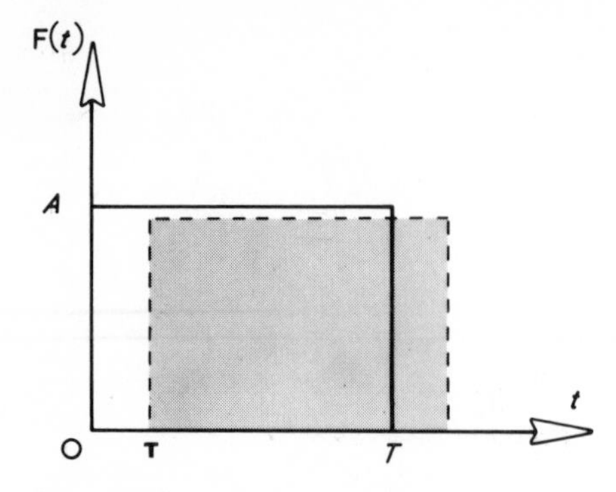

Fig. 14.4 Autocorrelating a rectangle

$$= \frac{A}{j2\pi f} \quad (1 - e^{-j2\pi f t})$$

$$= \frac{A}{2\pi f} \quad [\sin 2\pi Tf - j(1 - \cos 2\pi Tf)] \tag{14.24}$$

of which the squared modulus is

$$| F[f] |^2 = \frac{A^2}{2\pi^2 f^2} \quad (1 - \cos 2\pi Tf) \tag{14.25}$$

The cosine Fourier transform of $|F[f]|^2$ is

$$I = (A^2/2\pi^2) \int_0^\infty \frac{1 - \cos 2\pi Tf}{f^2} . \cos 2\pi \tau f \mathrm{d}f \tag{14.26}$$

$$= (A^2/2\pi^2) \int_0^\infty \frac{\cos 2\pi\tau f}{f^2} - \tfrac{1}{2} . \frac{\cos 2\pi\ (T + \tau)f}{f^2}$$

$$- \tfrac{1}{2} . \frac{\cos 2\pi(T - \tau)f}{f^2} \quad \mathrm{d}f \tag{14.26a}$$

Thus we have to integrate terms of the type

$$I_1 = \int_0^\infty \frac{\cos mf}{f^2} \quad \mathrm{d}f$$

where m is a constant. This is a very simple contour integral for which the method of integration is shown in Appendix 4. The answer is

$$I_1 = -\pi m/2 \tag{14.27}$$

Reverting now to Equation 14.26a and letting $m_1 = 2\pi\tau$, $m_2 = 2\pi T + m_1$ and $m_3 = 2\pi T - m_1$ we have

$$I = (A^2/2\pi^2)\left[I_1(m_1) - \tfrac{1}{2}I_1(m_2) - \tfrac{1}{2}I_1(m_3)\right]$$

$$= (A^2/2)(T-\tau) \qquad (14.28)$$

Now

$$C_a(\tau) = 2I/T$$

The factor 2 is the factor 2 before the integral 14.11d and accounts for the negative frequencies in the spectrum.

Finally $$C_a(\tau) = A^2(1 - |\tau/T|) \qquad (14.29)$$

14.1.6 The Wiener-Khintchine relation applied to a transient 'burst' of sine wave

The synchronising signal for the colour information in a colour television signal consists of a burst of about ten cycles of sine wave at the frequency of the colour-carrying subcarrier, namely $f_0 = 4.4$MHz. This burst is repeated at the beginning of each television line-scan and so its spectrum would be a line spectrum; however, we shall consider one burst of ten cycles in all time, so that the spectrum becomes a continuous one of density. Let

$$F(t) = A \sin 2\pi f_0 t \qquad 0 < t < 10/f_0 \qquad (14.30)$$

The autocorrelation function of $F(t)$ will be

$$C_a(\tau) = (A^2 f_0/10) \int_0^{10/f_0} \sin 2\pi f_0 t \cdot \sin 2\pi f_0 (t-\tau)dt \qquad (14.31)$$

After some trigonometric manipulation followed by quite simple integrations we have

$$C_a(\tau) = (A^2/2)\cos 2\pi f_0 \tau, \; -10/f_0 < \tau < 10/f_0 \qquad (14.31a)$$

The spectrum $F[f]$ is, of course,

$$F[f] = A \int_0^{10/f_0} \sin 2\pi f_0 t \cdot e^{-j2\pi f t}dt \qquad (14.32)$$

Again, some manipulation, followed by simple integrations followed, in turn, by further manipulation yield

$$F[\alpha] = (A/2\pi f_0) \quad (1 - e^{-j20\pi\alpha})/1 - \alpha^2) \qquad (14.32a)$$

where

$$\alpha = f/f_0$$

The square of the modulus of $F[\alpha]$ is

$$|F[\alpha]|^2 = (A^2/2\pi^2 f_0^2)(1 - \cos 20\pi\alpha) / (1-\alpha^2)^2 \qquad (14.33)$$

and the cosine Fourier transform is, omitting for the time being the factor $A^2/2\pi^2 f_0{}^2$,

$$I = f_0 \int_0^\infty \frac{(1 - \cos 20\pi\alpha) \cos 2\pi f_0 \tau\alpha \cdot d\alpha)}{(\alpha^2 - 1)^2} \tag{14.34}$$

The factor f_0, in front, arises from the fact that $df = f_0\ d\alpha$!

The integrand of I being an even function, we may write

$$I = (f_0/2)\left[\int_{-\infty}^{\infty} \frac{e^{jm\alpha}\ d\alpha}{(\alpha-1)^2\ (\alpha+1)^2} - \int_{-\infty}^{\infty} \frac{e^{jm_1\alpha}\ d\alpha}{(\alpha-1)^2\ (\alpha+1)^2}\right]$$

$$\text{where } m = 2\pi f_0 \tau, m_1 = 20\pi + 2\pi f_0 \tau \tag{14.34a}$$

in which the limits of integration are now $-\infty$ to $+\infty$ instead of 0 to ∞, and each cosine is replaced by the simple exponential since we know that the sine term being odd will be zero over the range covering all α's, both positive and negative (see Appendix 4). Proceeding as in Appendix 4 and noting that there are two double poles for $\alpha = \pm 1$, we have, for the residues of the two integrands,

$\frac{1}{4} e^{-jm}\ (jm + 1)$ and $\frac{1}{4} e^{jm}\ (jm - 1)$ for the first integrand

and

$\frac{1}{4} e^{-jm_1}\ (jm_1 + 1)$ and $\frac{1}{4} e^{jm_1}\ (jm_1 - 1)$ for the second intergrand.

Now adding 20π to m ($m_1 = 20\pi + m$) does not change the exponentials; thus

$$\begin{aligned} e^{\pm jm} &= e^{\pm jm \pm j20\pi} \\ &= e^{\pm j20\pi} \cdot e^{\pm jm} \end{aligned}$$

$$\begin{aligned} \text{and } e^{\pm j20\pi} &= \cos \pm 20\pi + j \sin \pm 20\pi \\ &= \cos 20\pi \\ &= 1 \end{aligned}$$

So the difference between the sums of the two residues of the first integrand and those of the second integrand in Equation 14.34a is, after some manipulation, $-j10\pi \cos m$. The integral I now becomes

$$\begin{aligned} I &= \frac{f_0}{2} \cdot \pi j \cdot (-j\,10\pi \cos m) \\ &= 5\pi^2 f_0 \end{aligned} \tag{14.34b}$$

We must now multiply I by $A^2/2\pi^2 f_0{}^2$, the factor we put aside after Equation 14.33, and we must also multiply by $2/T = 2f_0/10$ to obtain

$$C_a(\tau) = (A^2/2) \cos 2\pi f_0 \tau \qquad (14.35)$$

The integral I in Equation 14.34 is an extremely tricky one and contains many pitfalls for the unwary. If only one of the two cosines is replaced by the exponential, the answer becomes $I = 0$ which is, of course, false. I wonder if the last two Sections 14.1.5 and 14.1.6 have stimulated the reader's interest in contour integration!

The spectrum $F[\alpha]$ presents a good example of a relatively simple, but complex spectrum. We can re-write Equation 14.32a as

$$F[\alpha] = (A/2\pi f_0)\,(1 - \cos 20\pi\alpha + j \sin 20\pi\alpha)/(1 - \alpha^2) \qquad (14.32b)$$

The modulus is the square root of Equation 14.33 and this can easily be shown to be

$$|\,F[\alpha]\,| = A/\pi f_0 \quad \left| \frac{\sin 10\pi\alpha}{1 - \alpha^2} \right| \qquad (14.36)$$

whilst the phase angle, being the angle whose tangent is the ratio of the imaginary part of Equation 14.32b to the real part, is

$$\begin{aligned} &\arctan \left[\sin 20\pi\alpha/(1 - \cos 20\pi\alpha) \right] \\ &= \arctan \left[1/\tan 10\pi\alpha \right] \\ &= \pi/2 - 10\pi\alpha \qquad (14.37) \end{aligned}$$

Equation 14.36, without the coefficient $A/\pi f_0$, is shown in Fig. 14.5(b) and $|\,F[\alpha]\,|\,\underline{/\phi(\alpha)}$ is shown in polar co-ordinates in Fig. 14.5(a). For example [Fig. 14.5(a)], when $\alpha = 1.025$, $\phi = -9.75$, which is the same as $\pi/4$, and $|\,F[\alpha]\,| = 13.8$. The diagram starts in the centre, 0, with $\alpha = 0.8$ and, via the little arrowheads a, b, c, finishes at $\alpha = 1.2$. For $\alpha < 0.8$ or $\alpha > 1.2$ there would be smaller closed curves within the two shown in Fig. 14.5(a).

14.2 THE MEAN-POWER THEOREM

Consider a function of time $f(t)$ which is zero outside the time interval 0 to T; for example, a voltage waveform appearing across a resistance. Its autocorrelation function is

$$C_a(\tau) = 1/T \int_0^T f(t) \cdot f(t - \tau)\,dt \qquad (14.38)$$

which is, in effect, a re-write of Equation 14.5. Equation 14.6 leads us to

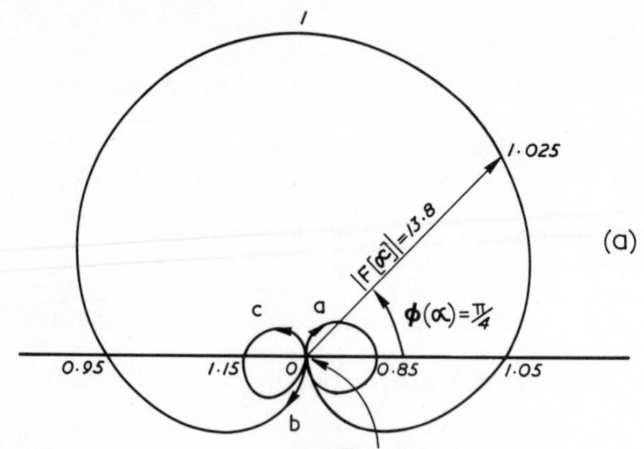

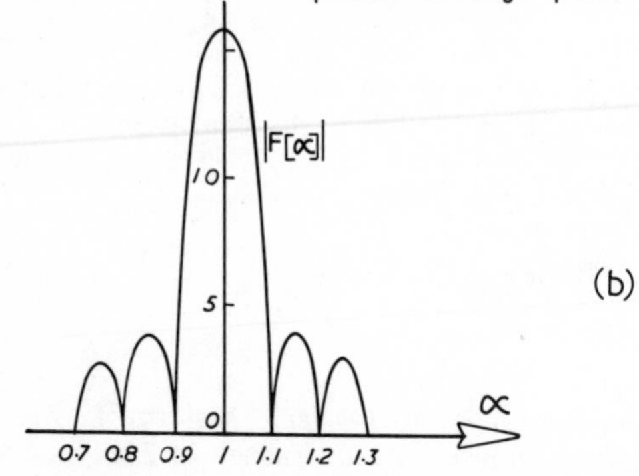

Fig. 14. 5 Complex spectrum of a 'burst' of sine wave

$$C_a(0) = 1/T \int_0^T f^2(t)dt \tag{14.39}$$

which tells us that $C_a(0)$ is the mean-square value of the function f. If, as suggested earlier, f is a voltage across a resistance, then $C_a(0)$ is proportional to the mean power.

Now we can also write, from Equation 14. 11b

$$C_a(\tau) = 1/T \int_{-\infty}^{\infty} |f[f]|^2 e^{j2\pi\tau f} df \tag{14.40}$$

and, again

$$C_a(0) = 1/T \int_{-\infty}^{\infty} |f[f]|^2 df \tag{14.41}$$

whence, Equating 14.41 to 14.39

$$\int_0^T f^2(t)dt = \int_{-\infty}^{\infty} |f[f]|^2 df \qquad (14.42)$$

We can replace the limits of integration of $\int_0^T f^2(t)dt$ by $-\infty$ and ∞, since if $f^2(t)$ has a finite area between these limits the integral will be finite, so finally

$$\int_{-\infty}^{\infty} f^2(t)dt = \int_{-\infty}^{\infty} |f[f]|^2 df \qquad (14.42a)$$

This justifies Equation 14.12 which assumes that the adding of squared spectral components is allowed.

Now when we are dealing with random noise we cannot know the function $f(t)$ since it will be changing unpredictably from moment to moment; but we often know its spectrum and therefore $|f[f]|^2$ and so we can easily calculate a figure proportional to the mean power by Equation 14.42a. Of course, we can also measure the mean power directly by putting a power-meter such as a heat sensitive device (thermocouple) in the circuit.

14.2.1 Typical example of use of power spectrum: the f.m./a.m. improvement factor[19]

The advantage of frequency modulation (f.m.) over amplitude modulation (a.m.) in sound broadcasting lies in the fact that for a given ratio of wanted carrier level to random noise level the receiver output signal-to-noise ratio can be a good deal greater for f.m. than for a.m. This advantage is offset by the requirement for a greater radio-frequency bandwidth in the f.m. case, and this means a wider 'channel allocation' and therefore the use of a higher-frequency band than the low-frequency (l.f.), medium-frequency (m.f.) and high-frequency (h.f.) bands used for a.m. sound broadcasting.

In order to examine the ways in which an a.m. receiver on the one hand and an f.m. receiver on the other hand will detect random noise, we must discuss the wanted signal formation and its demodulation in the receiver.

Basically, most analogue-type communication signals have the form

$$s = S \cos \phi \qquad (14.43)$$

that is, they have an amplitude $S(t)$ which may or may not be a function of time and a phase angle $\phi(t)$ which always will be a function of time. In the case of double-sideband amplitude modulation (d.s.b. a.m.) of the kind used in broadcasting, the amplitude becomes

$$\mathrm{S}(t) = \mathrm{A}(1 + m \sin at)$$

where 100 m is the percentage modulation by an assumed audio sinusoidal signal having an audio frequency of $a/2\pi$. A is the amplitude of the modulating signal which includes the d.c. term 1 whose function will appear presently. The phase ϕ, which, by definition is the integral of angular frequency, becomes

$$\phi = \int^{t} \omega_0 dt$$

$$= \omega_0 t$$

if the frequency of the carrier wave cos ϕ is a constant $\omega_0/2\pi$. Well, in d.s.b. a.m. the frequency of the carrier is always constant, so finally Equation 14. 43 becomes

$$\left.\begin{aligned} s &= A(1 + m \sin at) \cos \omega_0 t \\ &= A \cos \omega_0 t + Am \sin at . \cos \omega_0 t \\ &= A \cos \omega_0 t + (Am/2) \sin (\omega_0 + a) t \\ &\quad + (Am/2) \sin (\omega_0 - a) t \end{aligned}\right\} \quad (14.44)$$

The d.c. term 1 in the modulating signal gives rise to an unmodulated carrier $A \cos \omega_0 t$ which is necessary for the correct functioning of envelope detectors, which are cheap to make and reliable in operation. The d.s.b. a.m. demodulator or detector in the receiver simply converts the high-frequency factor $\cos \omega_0 t$ into a direct current, which is not passed on into the audio circuits of the receiver and retains the modulating signal $Am \sin at$ and this it does for all modulating signals having frequencies within the audio passband $\pm a_{max}/2\pi$. If random noise accompanies the signal s, then the detector will treat it in the same way as it does for the signal sidebands at frequencies $(\omega_0 + a)/2\pi$ and $(\omega_0 - a)/2\pi$. There is, however, a difference due to the fact that there is no coherent phase or timing relationship between a noise sideband at $(\omega_0 + a)/2\pi$ and its 'opposite number' at $(\omega_0 - a)/2\pi$. Whereas we can add the two sidebands constituting the second and third terms of the third Equation 14. 44 to obtain the second term of the second Equation 14. 44 we cannot do this in the case of two noise sidebands, because one will have an arbitrary and constantly changing unpredictable phase angle with respect to the other. We are, as regards noise, in the situation where we know the power spectrum, but not the waveform of the noise. In the case of a.m. demodulation, the power spectrum of the noise will take on the shape of the square of the spectrum of the noise waveform itself, which will be either that of the receiver intermediate-frequency (i.f.) circuits or that of the audio-frequency (a.f.) circuits, whichever is the narrower. Whichever is the narrower, we shall assume it to have a rectangular amplitude-versus-frequency band pass characteristic centred on ω_0 if we are dealing with a signal at i.f. or a band pass characteristic centred on zero frequency if we are dealing with a signal at a.f. A reminder is perhaps necessary here: a band pass characteristic centred on zero frequency is really the same as a low-pass characteristic in the range zero up to the cut-off frequency. I have, in this book, attempted to retain the notion of negative as well as positive frequencies which, it seems to me, avoids the difficulties so often encountered by engineers as to how upper and lower side-bands of a radio-frequency carrier both manage to appear in a low-pass characteristic after detection. If we retain the possibility of existence of negative frequencies, then detection may be regarded mathematically as a mere translation of a carrier from its frequency $\omega_0/2\pi$ to zero frequency where it becomes the d.c. component of the audio-, or, in general terms, the base-band signal.

For a d.s.b. a.m. receiver we shall assume that the i.f. bandwidth, being a band pass circuit, is exactly twice the a.f. bandwidth, being a low-pass circuit. Mathematically we shall regard the a.f. circuits as having an identical shape to that of the i.f. circuits, but centred around zero frequency.

Now assume that the spectrum of the noise appearing in the receiver is uniform and limited in frequency by the identical i.f. and a.f. characteristics. If the noise voltage density is N volts/Hz, then the mean noise power Pam will be proportional to the following integral

$$\left.\begin{aligned} Pam &\propto (1/2f_a) \int_{-f_a}^{f_a} N^2 \mathrm{d}f \text{ with } f_a = a_{max}/2\pi \\ \text{or } Pam &\propto N^2 \end{aligned}\right\} \quad (14.45)$$

because, with a rectangular-shaped spectrum and uniform noise density, N is constant.

We now consider the f.m. case. In Equation 14.43 we make S constant and equal to, say, A; but ϕ must now consist of a modulation of the frequency of the wanted carrier by the audio signal sin at. Thus, remembering that angular frequency is the rate of change of phase,

$$\mathrm{d}\phi/\mathrm{d}t = \omega_0 + \Delta\omega \sin at \quad (14.46)$$

where $\omega_0/2\pi$ is the 'rest' frequency of the carrier in the absence of modulation and $\Delta\omega/2\pi$ represents the maximum frequency deviation reached at the peak values of the audio modulating signal sin at when sin $at = \pm 1$. Equation 14.43 now becomes

$$\begin{aligned} s &= A \cos \left[\int^{t} (\omega_0 + \Delta\omega \sin at)\mathrm{d}t\right] \\ &= A \cos \left[\omega_0 t - (\Delta\omega/a) \cos at\right] \end{aligned} \quad (14.47)$$

so whilst the phase ϕ is increasing linearly with time, like a mains-driven electric clock, it has, superimposed upon it, a sinusoidal variation between the limits of $\pm\Delta\omega/a$ radians. The sidebands produced by a radio-frequency signal whose frequency is modulated by a sinusoidal audio signal are theoretically infinite in number instead of the two produced by a.m., but they fade away in strength as the difference in frequency between each of them and the at-rest frequency $\omega_0/2\pi$ increases. They may be calculated in strength and in frequency by means of Bessel functions, but we do not need to go so far here.

Now, how can we demodulate Equation 14.47? Let us take the derivative of it with respect to time.

$$ds/dt = -A(\omega_0 + \Delta\omega \sin at) \sin[\omega_0 t - (\Delta\omega/a) \cos at] \quad (14.48)$$

We obtain the sum of a constant-amplitude frequency-modulated wave $-A\omega_0 \sin[\omega_0 t - (\Delta\omega/a) \cos at]$ and a frequency-modulated wave which is amplitude modulated by the audio signal, thus $-A\Delta\omega \sin at \sin[\omega_0 t - (\Delta\omega/a) \cos at]$. If we pass the signal ds/dt into an envelope detector which is sensitive to amplitude variations, but insensitive to frequency variations, we shall obtain $-A\Delta\omega \sin at$ which is proportional to the original audio signal, plus a d.c. term equal to $A\omega_0$ from the constant-amplitude term in Equation 14.48. So the demodulator in an f.m. receiver first differentiates the i.f. signal and then passes the result to an ordinary a.m. envelope detector. The whole circuit is called a discriminator. Once again, as in the a.m. receiver, the envelope detector translates the carrier frequency $\omega_0/2\pi$ to zero so we can imagine the whole process occurring at positive and negative audio frequencies. In order to compare the noise level in the f.m. receiver output with that in the a.m. receiver output, we must equalise the gains in the two hypothetical receivers. We must ensure that the audio signal output from the detector in the discriminator equals that from the a.m. receiver's detector, so for a maximum angular frequency deviation $\Delta\omega$ (maximum modulation depth) the gain of the f.m. receiver must be such that the noise level is N volts/Hz, (at $\omega_0 + \Delta\omega$ in the i.f. part of the discriminator or simply $\Delta\omega$ at baseband or a.f.), the same as the a.m.-receiver noise. The reason I have had to equalise the gains for a signal at maximum deviation in the f.m. case is one of convenience of explanation—I could have done it at any deviation—because the noise spectrum in the f.m. case is not uniform, but triangular, as we shall see.

Now, from rule (i) in Chapter 9, we know that the spectrum of the differential operator d/dt is $p = j\omega = j2\pi f$, so the time-differentiating property of the discriminator translates itself in the spectrum world into a multiplication by $j2\pi f$ of modulus $2\pi f$ and phase angle $\pi/2$. The latter property we may neglect, since it merely changes a sine into a cosine or vice-versa, (with a change of sign). Thus, after passage through the differentiating portion of the discriminator, the uniform noise spectrum N within the i.f. band $\omega_0 - \Delta\omega < \omega < \omega_0 + \Delta\omega$ has the form shown in Fig. 14.6(a). The slope K of the discriminator characteristic is in the hands of the designer and need not concern us. After envelope detection the noise spectrum is as shown in Fig. 14.6(b). It is easy to see that the spectrum is a linear function of frequency deviation ω up to the maximum $\Delta\omega$ for which the system is designed;

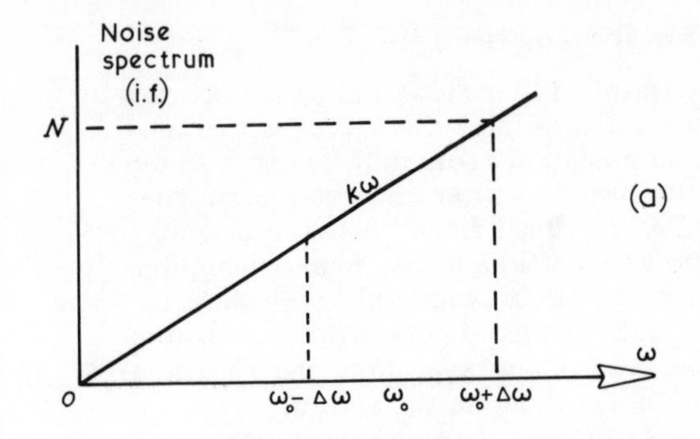

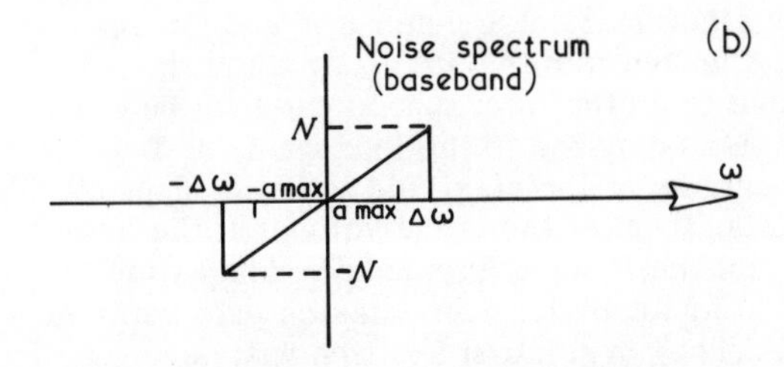

Fig. 14. 6 Spectrum of random noise in a frequency-modulation receiver

that is $N\omega/\Delta\omega$. The power spectrum is proportional to $(N\omega/\Delta\omega)^2$ or $(Nf/\Delta f)^2$ and the relative amount of noise power which can get through the audio circuits of bandwidth $\pm a_{\mathrm{max}}$ or f_{a} is

$$\left.\begin{aligned} P_{\mathrm{fm}} &\propto (1/2f_{\mathrm{a}}) \int_{f_a}^{f_a} N^2 f^2/(\Delta f)^2 \, \mathrm{d}f \\ &\propto (1/2f_a) N^2/(\Delta f)^2 \int_{f_a}^{f_a} f^2 \mathrm{d}f \\ &\propto (Nf_{\mathrm{a}}/\Delta f\sqrt{3})^2 \end{aligned}\right\} \qquad (14.49)$$

The ratio of a.m. noise power to f.m. noise power is therefore $(\Delta f \cdot \sqrt{3}/f_{\mathrm{a}})^2$ and the ratio of f.m. signal-to-noise ratio to a.m. signal-to-noise ratio is $\sqrt{3} \cdot \Delta f/f_a$. In a typical case $\Delta f = 75$ kHz and $f_{\mathrm{a}} =$ 15 kHz (say) giving an f.m.-to-a.m. improvement of 8.7, or 19dB. The use of a technique known as pre-emphasis of the transmitted signal followed by de-emphasis in the receiver adds a few decibels to the above improvement.

14.2.2 Use of correlation for acoustic tests[20]

It is important to keep to a minimum the leakage of sound between a broadcasting studio and other parts of the building; in particular, sound from the loudspeaker in a control room adjacent to a studio must not leak excessively into the studio itself where the microphone(s) could pick it up and cause 'howl round' between the two areas. In trying to locate the weakest link in the sound insulation between a control room and its studio it is helpful to measure the time delays appertaining to each 'leaky' transmission path between the source of sound—in this case, the loudspeaker in the control room—and the sink—in this case, the microphone in the studio.

Without going into the complexities of the measurement procedure, suffice it to say that a successful technique is to emit random acoustic noise from the control-room loudspeaker by feeding it from a uniform-spectrum random electrical noise generator and picking-up the acoustic noise that has leaked into the studio by using the studio microphone. The output from the microphone can then be correlated with the electrical noise being fed to the loudspeaker. Let $F[f] = N$ volts/Hz be the spectrum of the noise fed to the loudspeaker and let $F[f]e^{-j2\pi f\theta}$ be the spectrum of the noise voltage at the output terminals of the studio microphone; θ being an unknown delay time which will represent a particular acoustic-transmission path between loudspeaker and microphone. The correlation function will be

$$C(\tau, \theta) = \int_{-f_a}^{f_a} |F[f]|^2 e^{-j2\pi f\theta} \cdot e^{j2\pi\tau f} df$$

or

$$C(\tau - \theta) = \int_{-f_a}^{f_a} |F[f]|^2 e^{j2\pi(\tau - \theta)f} df \qquad (14.50)$$

Once again, the maximum of this function occurs at C(o) when $\tau = \theta$; so the position of the maximum is that for which $\theta = \tau$, whence we have found θ.

Appendix 1

An example of a Fourier-series expansion

Consider the function

$$h(t) = h_1(t) + h_2(t)$$

where

$$h_1(t) = -1 \quad \text{for } -T/2 < t < 0$$

$$= 1 \quad \text{for } 0 < t < T/2$$

and

$$h_2(t) = -\sin \omega t \quad -T/2 < t < 0$$

$$= \sin \omega t \quad 0 < t < T/2$$

where $\omega = 2\pi/T$. The function h(t) is shown in Fig. 10.11.

Now, if we re-write Equation 10.37 in Section 10.4 as

$$h(t) = 1/T \int_{-T/2}^{T/2} h(\tau) e^{jn\omega\tau} d\tau \sum_{n=-\infty}^{\infty} e^{-jn\omega t} \qquad (10.37a)$$

we have, for $h_1(t)$,

$$h_1(t) = (1/T)\left[-\int_{-T/2}^{0} e^{jn\omega\tau} d\tau + \int_{0}^{T/2} e^{jn\omega\tau} d\tau \right] \sum_{n=-\infty}^{\infty} e^{jn\omega t}$$

$$= 1/(j2\pi n)\left[e^{-jn\pi} - e^0 + e^{jn\pi} - e^0 \right] \sum_{n=-\infty}^{\infty} e^{-jn\omega t}$$

Using de Moivre's theorem

$$h_1(t) = 1/(j\pi n)(\cos n\pi - 1) \sum_{n=-\infty}^{\infty} e^{-jn\omega t}$$

and again using de Moivre's theorem on the exponential function in the summation sign

$$h_1(t) = \sum_{n=-\infty}^{\infty} \left(\frac{\cos n\pi - 1}{jn\pi} \cos n\omega t - \frac{\cos n\pi - 1}{n\pi} \sin n\omega t \right)$$

The first term is odd in n and is therefore zero (it is zero also when $n = 0$), when the summation is completed. The second term, being even in n and also being zero when $n = 0$, we have

$$h_1(t) = (2/\pi) \sum_{n=1}^{\infty} \frac{1 - \cos n\pi}{n} \sin n\omega t \qquad \text{(A. 1)}$$

Calculation of $h_2(t)$ is similar; thus

$$h_2(t) = (1/T) \left[- \int_{-T/2}^{0} \sin \omega\tau \cdot e^{jn\omega\tau} d\tau + \int_{0}^{T/2} \sin \omega\tau \cdot e^{jn\omega\tau} d\tau \right] \sum_{n=-\infty}^{\infty} e^{-jn\omega\tau}$$

$$= (1/T) \left[\int_{0}^{-T/2} \frac{e^{j\omega(n+1)\tau} - e^{j\omega(n-1)\tau}}{2j} d\tau \right.$$

$$\left. + \int_{0}^{T/2} \frac{e^{j\omega(n+1)\tau} - e^{j\omega(n-1)\tau}}{2j} d\tau \right] \sum_{n=-\infty}^{\infty} e^{-jn\omega t}$$

$$= -(1/4) \left[\frac{e^{-j\pi(n+1)} - e^0}{n+1} - \frac{e^{-j\pi(n-1)} - e^0}{n-1} \right.$$

$$\left. + \frac{e^{j\pi(n+1)} - e^0}{n+1} - \frac{e^{j\pi(n-1)} - e^0}{n-1} \right] \sum_{n=-\infty}^{\infty} e^{-jn\omega t}$$

Of course, $e^{\pm j\pi} = -1$, so using de Moivre's theorem again,

$$h_2(t) = (1/2\pi) \left[\frac{1 + \cos n\pi}{n+1} - \frac{1 + \cos n\pi}{n-1} \right] \sum_{n=-\infty}^{\infty} e^{-jn\omega t}$$

Now we put the expression in square brackets inside the summation and multiply it by the exponential and obtain

$$h_2(t) = (1/\pi) \sum_{n=-\infty}^{\infty} \left(\frac{1 + \cos n\pi}{1 - n^2} \cos n\omega t - \frac{1 + \cos n\pi}{1 - n^2} \sin n\omega t \right)$$

Thr first term is even and the second term is odd and when $n = 0$ the first term equals 2, thus

$$h_2(t) = (2/\pi)\left[1 + \sum_{n=1}^{\infty} \frac{1 + \cos n\pi}{1 - n^2} \cos n\omega t\right] \quad \text{(A. 2)}$$

Adding Equation A. 1 to A. 2 we have, finally

$$h(t) = (2/\pi)\left[1 + \sum_{n=1}^{\infty} \frac{1 + \cos n\pi}{1 - n^2} \cos n\omega t + \frac{1 - \cos n\pi}{n} \sin n\omega t\right] \quad \text{(A. 3)}$$

or

$$h(t) = (2/\pi)(1 + 2 \sin \omega t - (2/3) \cos 2\omega t + (2/3) \sin 3\omega t - (2/15) \cos 4\omega t + (2/5) \sin 5\omega t - (2/35) \cos 6\omega t + (2/7) \sin 7\omega t - (2/63) \cos 8\omega t + 100) \quad \text{(A. 3a)}$$

Appendix 2

Algebraic Long Division: an example

		$-z^5$	$0z^6$	z^7	$2z^8$	$3z^9$	$3z^{10}$	$= f[z]/h[z] \doteqdot f(x)*/h(x)$			
$h[z] =$	$z^3 \quad 3z^4 \quad 5z^5 \quad 2z^6$	$-z^8$	$-3z^9$	$-4z^{10}$	$3z^{11}$	$14z^{12}$	$24z^{13}$	$28z^{14}$	$21z^{15}$	$6z^{16}$	$= f[z]$
		$-z^8$	$-3z^9$	$-5z^{10}$	$-2z^{11}$						
		0	0	z^{10}	$5z^{11}$	$14z^{12}$	$\leftarrow$D				
				0	0	0	0	because having 'brought down' the $14z^{12}$ there are still not enough terms for the difference D to be divided by the divisor, $(z^3\ 3z^4\ 5z^5\ 2z^6)$			
				z^{10}	$5z^{11}$	$14z^{12}$	$24z^{13}$				
				z^{10}	$3z^{11}$	$5z^{12}$	$2z^{13}$				
				0	$2z^{11}$	$9z^{12}$	$22z^{13}$	$28z^{14}$			
					$2z^{11}$	$6z^{12}$	$10z^{13}$	$4z^{14}$			
					0	$3z^{12}$	$12z^{13}$	$24z^{14}$	$21z^{15}$		
						$3z^{12}$	$9z^{13}$	$15z^{14}$	$6z^{15}$		
						0	$3z^{13}$	$9z^{14}$	$15z^{15}$	$6z^{16}$	
							$3z^{13}$	$9z^{14}$	$15z^{15}$	$6z^{16}$	
					remainder:		0	0	0	0	

Appendix 3

Arithmetic Convolution Division: same example as in Appendix 2

		3	4	5	6	7	8	9	10	11	12	13	14	15	16	←	x and column number
							−1	−3	−4	3	14	24	28	21	6	←	f(x)
		1	3	5	2											←	h(x)
				−1	0	1	2	3	3	$= f(x) * / h(x) \doteqdot f[z]/h[z]$							
h(x)	→	1	3	5	2		−1	−3	−4	3	14	24	28	21	6	f(x)	
							−1	−3	−5	−2							
							0	0	1	5	14						
									0	0	0	0					
									1	5	14	24					
									1	3	5	2					
									0	2	9	22	28				
										2	6	10	4				
										0	3	12	24	21			
											3	9	15	6			
												3	9	15	6		
												3	9	15	6		
											remainder:	0	0	0	0		

Note that the quotient's first term (−1) is in column 5, because it is the quotient of the (−1) of the dividend, in column 8 and the (1) of the divisor, in column 3; and $8 - 3 = 5$

Appendix 4

To integrate[10]

$$I_1 = \int_0^\infty \frac{\cos mf}{f^2}\, df$$

we note, first that the integrand is an even function, so we may write

$$I_1 = (1/2) \int_{-\infty}^{\infty} \frac{e^{jmf}}{f^2}\, df$$

since the term $j \sin mf$ in the expanded exponential will integrate to zero, because of the symmetrical limits of integration. Second the integrand has a pole (denominator zero) at $f = 0$. Third, the pole is a double one, since the denominator is f^2. Fourth, the integrand tends to zero as f tends to infinity.

We therefore place the integrand in the complex plane and to avoid the pole, the path of integration becomes as shown in Fig. A. 1.

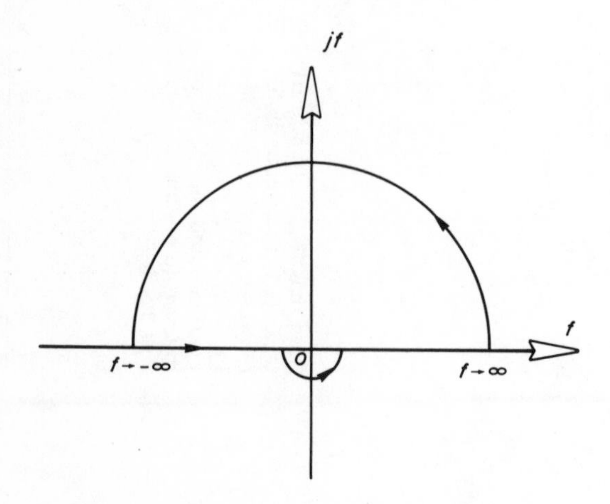

Fig. A. 1. Path of integration for pole at zero

The integrand is zero along the big semicircle and the residue of the double-poled integrand may be obtained by taking the derivative with respect to f of the integrand after having 'cleared' the pole by multiplication by the denominator f^2 and subsequently replacing f by its value at the pole, namely zero. Thus

$$\text{residue} = \frac{d}{df}(f^2 e^{jmf}/f^2)_{(f=0)}$$

$$= (jm\, e^{jfm})_{(f=0)}$$

$$= jm$$

Because, in tracing the path shown in the figure, we only half circumnavigate the pole, we multiply the residue by πj instead of $2\pi j$ thus, finally

$$I_1 = (1/2).jm.j\pi$$
$$= -\pi m/2$$

References

1 Craddock, J. M., *Statistics in the Computer Age*, p. 57, E.U.P., London (1969).
2 Prosser, Allnatt, and Lewis, 'Quality grading of impaired television pictures', *Proc. IEE*, 1964, vol 111 no. 3.
3 Palin Elderton, W., *Frequency Curves and Correlation*, 2nd edn. Charles and Edwin Layton, London (1927).
4 Woodward, P. M., *Probability and Information Theory with Applications to Radar*, 2nd edn. Pergamon Press, Oxford (1964).
5 Peirce, B. O. *A Short Table of Integrals*, 3rd edn. Ginn & Co., London (1929).
6 Dobesch and Sulanke, *Zeitfunktionen*, VEB Verlag Technik, Berlin (1964).
7 Carson, J. R., *Electric Circuit Theory and the Operational Calculus*, McGraw Hill, New York and London (1926).
8 Campbell, George A., and Foster, Ronald M., *Fourier Integrals for Practical Applications*, Van Nostrand, New York, Toronto and London (1948).
9 *The Heaviside Centenary Volume*, The Institution of Electrical Engineers, London (1950).
10 Angot, Andre, *Complements de Mathématiques*, Editions de la Revue d' Optique, Paris (1949).
11 Titchmarch, E. C., *Introduction to the Theory of Fourier Integrals*, 2nd edn., Oxford University Press (1948).
12 Whittaker, E. T., and Watson, G. N., *A Course of Modern Analysis*, Cambridge University Press (1965). (A very fundamental treatise).
13 Maurice, R. D. A. and Rout, E. R., 'Characteristics of Flywheel Synchronising Circuits in, Television Receivers, *Electronic Engineering*, 1962, February, p. 77.
14 *Synchronizing and Line Oscillator Circuits for an Experimental 625-line Receiver*, Mullard Technical Communications, 1958, 4/80.
15 Janke E. and Emde F., *Tables of Functions*, Dover, New York (1943).
16 I am indebted to my colleague G. D. Monteath of the BBC Research Department for permission to use his methods described in *BBC Engineering Department Monograph*, 1962, no. 45, December.
17 Gibson, W. G., and Schroeder, A. C., 'A vertical aperture equalizer for television', *J.S.M.P.T.E.*, 1960, vol. 69, no. 6, June.
18 Tustin, A., 'A method of analyzing the behaviour of linear systems in terms of time series', *Journal IEE*, 1947, vol. 94, 111A, no. 1 pp. 130-142.

19 Maurice, R. D. A. 'VHF broadcasting', *Electronic & Radio Engineer,* 1957, August, p. 300 et seq.
20 Burd, A. N. 'Correlation techniques in studio testing', *Radio & Electronic Engineer,* 1964, May, p. 387 et seq.

Index